PAST TIME, PAST PLACE

GIS for History

Edited by Anne Kelly Knowles

ESRI Press
REDLANDS, CALIFORNIA

ESRI
Past Time, Past Place: GIS for History
ISBN 1-58948-032-5

First printing March 2002.

Printed in the United States of America.

Library of Congress Cataloging-in-Publication Data
Past time, past place : GIS for history / Anne Kelly Knowles, editor.
p. cm.
ISBN 1-58948-032-5 (pbk.)
1. Geographic information systems. 2. Historical geography-Methodology.
I. Knowles, Anne Kelly.
G70.212.P38 2002
910'.285—dc21 2002002688

Published by ESRI, 380 New York Street, Redlands, California 92373-8100.

Books from ESRI Press are available to resellers worldwide through Independent Publishers Group (IPG). For information on volume discounts, or to place an order, call IPG at 1-800-888-4741 in the United States, or at 312-337-0747 outside the United States.

Contents

PREFACE *vii*
Myron P. Gutmann

INTRODUCING HISTORICAL GIS *xi*
Anne Kelly Knowles

1 HISTORICAL MAPS IN GIS 1
David Rumsey and Meredith Williams

2 TEACHING THE SALEM WITCH TRIALS 19
Benjamin C. Ray

3 SIMILARITY AND DIFFERENCE IN THE ANTEBELLUM NORTH AND SOUTH 35
Aaron C. Sheehan-Dean

4 TELLING CIVIL WAR BATTLEFIELD STORIES WITH GIS 51
David W. Lowe

5 IMMIGRATION, ETHNICITY, AND RACE IN METROPOLITAN NEW YORK, 1900–2000 65
Andrew A. Beveridge

6 REDLINING IN PHILADELPHIA 79
Amy Hillier

7 CAUSES OF THE DUST BOWL 93
Geoff Cunfer

8 AGRICULTURAL HISTORY WITH GIS 105
Alastair W. Pearson and Peter Collier

9 MAPPING BRITISH POPULATION HISTORY 117
Ian N. Gregory and Humphrey R. Southall

10 GIS in Archaeology 131
Trevor M. Harris

11 Mapping the Ancient World 145
Tom Elliott and Richard Talbert

12 The Electronic Cultural Atlas Initiative and the North American Religion Atlas 163
Lewis R. Lancaster and David J. Bodenhamer

Glossary of GIS Terms 179

About the Contributors 199

Preface

From historians who wrote long ago to those who write today, geographical location has been an abiding theme. Put another way, many historians are part geographer, as interested in "where" events occurred as they are in "who," "how," and "why." We—and I happily count myself among them—see maps as vital tools in understanding human behavior.

Over the past fifteen years the work of map-loving historians has been made easier and more exciting by the availability of personal computers and geographic information system (GIS) software. Beginning with simple but powerful tools that allowed researchers to bring existing maps within the computer environment for easy reproduction, transformation, and presentation, or to make maps using computer cartography, the software and the computers on which it is used have become more and more sophisticated. It is now possible to devise dynamic views of past experience in the form of animated series of map images that can be started and stopped as "time" progresses to show conditions at any given moment. It is also possible to bring together multiple maps in a single image, and to represent and analyze information as multiple layers in a single map document. These capabilities could well make map lovers of a new generation of computer-literate historians, and bring computing expertise to future generations of map lovers.

The chapters of this book demonstrate the range of possibilities that become available when history teachers and researchers use GIS to inform and enhance their work. Just as increasing computational capacity gave impetus to the quantitative revolution of the 1960s and 1970s, GIS is putting spatial analysis within reach of virtually any user. It can reveal the spatial patterns of people, things, even transient events, whether the item recorded is an archaeological artifact, a house, a block, a parish, a county, or even a dust storm. Beyond location, GIS allows the researcher to record and retrieve many of the characteristics of the item identified. The chapters show us photographs of the artifact, characteristics of the individual residents of the house, enumerations of the population of the block, parish, or county, and the intensity of the dust storm. Students and scholars can probe the location and the characteristics simultaneously to tell an enriched story about the past.

In one of its simpler forms, creating a GIS allows us to define and map spatial polygons that reflect a location on the ground. These boundaries are commonly institutional jurisdictions, such as a census tract, town, county, state, or province. We map the outline of the jurisdiction, and then store all manner of information about that territory: how it is governed, its population by age and sex, number of horses, cows, or farms, number of businesses and workers, average elevation, or total monthly precipitation. With this kind of GIS we can map the characteristics of the blocks of space represented by the polygons, showing different characteristics through color and symbol. This adds a whole new dimension to conventional quantitative and qualitative analysis.

Nor need we restrict our GIS to polygons and to a visual comparison of their attributes. The chapters in this volume are rich in examples of GIS elements that are points (houses, churches, archaeological artifact locations) and lines (battlefield trenches, rivers and streams). Point and line locations and data can be overlaid on polygon data to allow us to visualize where members of churches live ("The Electronic Cultural Atlas Initiative and the North American Religion Atlas"), or where mortgages were made ("Redlining in Philadelphia"). Moreover, they can be analyzed with a large suite of spatial tools, including the drawing of zones of significant distance ("buffers") around houses, churches, or towns ("Similarity and Difference in the Antebellum North and South") that allow students and researchers to understand how distance from a point or polygon, or travel time along a road, shaped past experiences and conditions.

GIS enriches both qualitative and quantitative approaches to history. For those of us who "see" relationships in the visual dimensions of a map, GIS gives us the opportunity to quantify distances, directions, and attributes, and then study their statistical characteristics. For those of us for whom quantitative and statistical relationships are an accustomed way of seeing the past, the opportunity to "see" the space of the past is extremely valuable. Studies that perform activities like the mapping of the residuals from regression analysis demonstrate that the statistical and the visual can play a coupled role.[1] It is these combinations of new ways to see the past that has enabled students, teachers, and researchers to think differently about the past.

Many of the chapters in this volume show how historical maps can be brought into the GIS, even when they are primitive or known to be inaccurate. The authors have taken older manuscript and printed maps, and then rectified their placement in the GIS by specifying known locations, often by going into the field with a Global Positioning System (GPS) device. Once specific locations on the historical map have been pinned down, the rest of the map can be adjusted computationally to establish relatively accurate locations for the map in its entirety. Where researchers desire three-dimensional effects, this process of fitting historical maps to known locations allows an older map originally prepared on a flat piece of paper to be "draped" over a three-dimensional spatial model, in order to allow visualization, as was the case in "Historical Maps in GIS" and "Telling Civil War Battlefield Stories with GIS."

The visual and analytic tools of GIS go well beyond merely illustrating history. The teaching and research revealed in the chapters of this book show just how exciting are the questions that can be answered. This preface can't summarize them all, but even a short list suggests how useful and stimulating GIS can be in serious historical inquiry:

- Did where people live determine whether they were accused or accusers in Salem?
- Does detailed information about the location of businesses, farms, and voters on a local level reveal the causes of the Civil War?

1. See, for example, M. P. Gutmann and C. G. Sample, "Land, Climate, and Settlement on the Texas Frontier," *Southwestern Historical Quarterly* 99 (1995): 137–72.

- How can we in the early twenty-first century visualize battlefield conditions of the Civil War?
- Where did migrants and members of different ethnic groups live in New York in the twentieth century?
- What do the spatial distribution of houses and economic conditions tell us about the ways that New Deal policies may have led to the creation or confirmation of redlining (restrictive loan policies) in American cities?
- Does the overlapping spatial distribution of farming and dust storms tell us how the Dust Bowl of the 1930s was created?
- Did the largest landowners in Wales really make the most innovations in agriculture during the Agricultural Revolution?

More than sixty years ago, influential works by French historians, such as Marc Bloch's *French Rural History: An Essay on its Basic Characteristics,* brought a new concern for the geographic detail of the land into the teaching and study of history. These innovators were part of a wave of new ideas about history that brought sophisticated concepts from the social sciences and eventually from literature to the study of the past. In the years after the second world war these new ideas took exquisite form in Fernand Braudel's *The Mediterranean and the Mediterranean World in the Era of Philip II* and in a broad range of studies on both sides of the Atlantic that brought innovations from many other fields of study to history. It is easy to predict that when we recollect the development of history at the end of the twentieth and the beginning of the twenty-first century, the introduction of GIS to research and teaching about the past will be one of the signs of the successful continuation, and reinvigoration, of that tradition.

Myron P. Gutmann
Ann Arbor
21 December 2001

INTRODUCING HISTORICAL GIS

Anne Kelly Knowles

This book is the first collection of case studies applying geographic information systems (GIS) to the study of history. It announces the emergence of a compelling methodology—historical GIS—that is generating interest among historical scholars working in history, geography, sociology, anthropology, archaeology, religious studies, and many other disciplines in the social sciences and humanities. Over the past seven years or so, developments in historical GIS have cohered in a set of common techniques, sources, and issues. It is time for those working with the approach to share what they have learned with their fellow scholars, teachers, and students.

Each chapter discusses how geographic information and GIS technology can facilitate historical inquiry and what GIS enables one to do that previously would have been much harder or impossible to do. While the contributors highlight the advantages of using GIS, they also explain the difficulties of applying GIS to historical sources and to the kinds of problems people can expect in their first encounters with the technology. For many scholars working in disciplines other than geography, GIS is a new and daunting tool. For many geographers, particularly those who are adept at using GIS, the past itself is unknown terrain and historiography an unfamiliar framework for research questions. We cannot hope to fill all the gaps for all readers in a single volume. Our aim is to share the early fruits of an exciting new approach and to point toward future developments that will give historical GIS a strong and lasting foundation.

THE CONVERGENCE OF HISTORY AND GEOGRAPHY IN HISTORICAL GIS

The old saw that history is the study of when, geography the study of where, has some truth to it, at least as the two fields have been taught at most colleges and universities in North America. Traditionally, historians acknowledged geography mainly as the physical landscape, meriting little more than a few pages of artful description, or as location in its simplest sense, noted when necessary in "locator" maps. In geography over the past twenty years, the study of contemporary phenomena eclipsed interest in history, to the point that today historical geography is not taught in some geography departments in the United States. But there have always been historians fascinated by maps and geography just as there remains a strong core of historical interest in geography. Recent indications of renewed interest in empirical approaches to history, after a generation of theoretical development, include the emergence of specifically geographical fields of study such as regional history and environmental history. At the same time, GIS experts are grappling with the complexities of analyzing and representing change over time. Historical GIS has thus become a meeting ground for historians, geographers, geographic information scientists, and others who previously had little contact. This convergence of interests and approaches for solving research problems recasts the old adage in a new light. Geography is the study of spatial differentiation, history the study of temporal differentiation. Historical GIS provides the tools to combine them to study patterns of change over space and time.

Using GIS intelligently requires a grounding in geographical knowledge. Applying the technology to history requires knowing how to contextualize and interpret historical sources. The collaborative teams tackling historical questions with GIS, including many of the projects represented in this book, testify to how much we have to learn from one another as we develop this truly interdisciplinary approach. The ability of GIS to integrate, analyze, and visually represent spatially referenced information is inspiring historians to combine sources in new ways, to make geographical context an explicit part of their analysis, to reexamine familiar evidence, and to challenge long-standing historical interpretations. Historical GIS is proving increasingly valuable as a research method, a framework for digital archives, and a means of bringing a geographical sensibility to our view of history.

ORGANIZATION AND AUDIENCE

The diverse subjects of the twelve essays in this book indicate the tremendous range of historical topics now being explored with the aid of GIS and related spatial technologies. Chapter 1, "Historical Maps in GIS," explains how historical maps provide a foundation for doing historical GIS. The six following chapters offer new perspectives on key themes in U.S. history, from the Salem witch trials and the Civil War to the changing human mosaic of New York City and the causes of the Dust Bowl. These essays can be used as a focused geographical supplement to the U.S. History survey course. They can also be used in upper-level methods courses in history, anthropology, sociology, geography, and other disciplines as examples of how researchers are applying GIS to specific research questions at the local and regional scale. Chapters 8, 9, and 10 deal with more advanced techniques, such as spatial statistics and 3-D modeling. The last two chapters look toward the future of GIS as a framework for organizing digital historical archives and for coordinating and sharing historical scholarship globally. They will be of special interest to scholars who are planning multiscale, interdisciplinary projects involving GIS.

Past Time, Past Place is intended for people who are either fairly new to GIS or are just beginning to apply the technology to historical questions. Contributors therefore explain basic GIS concepts and methods in context, and briefly summarize the historiographic debates relevant to their case studies. GIS terms, highlighted in italics, are defined in the text and in the glossary at the end of the book. Readers interested in pursuing technical matters or historical topics in greater depth can consult the publications listed in the "Further reading" section at the end of each chapter. The further reading lists also include references to online sources of digital historical data and teaching resources suitable for historical GIS. For information on training courses and college-level educational opportunities for learning GIS, we recommend ESRI's information clearinghouse (www.gis.com) as a point of entry. It provides links to undergraduate and graduate programs, training workshops, conferences, professional organizations, textbooks, and the exploding specialist literatures on geographic information systems and geographic information science. Ian N. Gregory's *A Place in History: A Guide to Using GIS in Historical Research,*[1] a more technical introduction to historical GIS, is a particularly good complement to this text.

THE NATURE OF GEOGRAPHIC INFORMATION AND GIS

Geographic information is qualitative or quantitative information that can be located on the surface of the earth. People are often surprised to realize that most of the information they use every day has a geographic component, from addresses in a telephone book to the inventory systems used by retail firms. Many of the sources familiar to historians contain geographic information, but much of it has gone unrecognized or unused. In addition to the necessary quality of location, geographic information possesses other characteristics that distinguish it from other kinds of information. It is multidimensional, requiring at least two coordinates to define location and at least three if the item being located in space is also being located in time. It is often voluminous, because capturing the geographic extent of a large area and its features can easily generate gigabytes of data. It is often best represented in graphic form, particularly in maps, rather than in tables, charts, or text. And analyzing geographic information can require methods that are new to many in the social sciences and humanities.[2]

A *geographic information system* (GIS) digitally links locations and their attributes so that they can be displayed in maps and analyzed, whether by their geographical characteristics, such as location, distance, proximity, density, and dispersal, or by their attributes, such as social, economic, and physical characteristics. In *vector* GIS, spatial features exist as points, lines, and polygons that represent, for example, cities, roads, and states. Their attributes might be population, traffic

1. Ian N. Gregory, *A Place in History: A Guide to Using GIS in Historical Research* (Oxford: Oxbow Books, 2002). Gregory's *Guide* is also available online at hds.essex.ac.uk/g2gp/gis/index.asp.

2. Paul A. Longley, Michael F. Goodchild, David J. Maguire, and David W. Rhind, *Geographic Information Systems and Science* (Chichester: Wiley, 2000), 6.

volume, and industrial output. In *raster* GIS, space is modeled as a continuous surface made up of cells called *pixels,* each of which is assigned one or more values representing attributes such as elevation, soil type, vegetation, or temperature. Satellite images, such as the shifting weather patterns shown on the nightly news, are examples of raster GIS. Vector GIS is most often used with social and economic data.

The organizing principle in all GIS is location, as encoded in geographical *coordinates* such as degrees of latitude and longitude. Location is what gives GIS its unique powers to integrate and analyze data. Once points, lines, polygons, and pixel cells are assigned coordinates, a process called *georeferencing,* one can combine them in layers with any other features or surfaces that are registered to the same location. Layering can be done for simple visual comparison. It is also the basis for methods of *spatial analysis* such as *overlay,* in which the GIS integrates and analyzes two or more layers of geographic information. In addition to enabling certain analytical functions, location makes GIS an extremely useful framework for compiling and indexing information. Libraries and archives around the world are developing new systems for structuring research collections, books, film and sound archives, and other resources according to location. New kinds of geographical search engines, called *gazetteers,* are being designed that will enable users to search for the many kinds of information related to a given place. This new geographical paradigm for information storage and retrieval is spurring further advances in GIS technology and creating new opportunities for place-based historical research.

Many newcomers to GIS associate the technology with making maps. One certainly can use GIS to make maps. One of the technology's most appealing advantages is that, once spatial and attribute data is correctly entered into the system, a GIS can almost instantly generate maps in answer to queries, and can do so as easily for a very large data set as for a very small one. The ease of mapping in GIS removes the technical obstacles that formerly limited most scholars to mapping only their final results, and even then only with the help of skilled cartographers. While compelling, handsome, thoroughly convincing cartography remains a fine art, GIS has made it possible for those without cartographic training to explore the geographical patterns in data through on-screen visualization.

Iterative mapping—producing many maps in answer to slightly varied queries—becomes part of the research process in historical GIS. It also cultivates geographical thinking and helps one formulate geographical questions, which in the long run may work greater change in historical scholarship than will the technology's facility as a cartographic tool.

The statistical capabilities of GIS are rapidly improving. Spatial statistics are influencing developments in the "new economic geography," to name just one of the fields applying them to social science research. Traditional statistical measures assume that individual observations are independent of one another. Spatial statistics are based on the very different assumption, sometimes called The First Law of Geography, that everything is related to everything else, but near things are more related than those that are far apart. The basic statistical measure that expresses this law is called *spatial autocorrelation.* If features close to each other have similar attributes, the pattern is said to show positive spatial autocorrelation. If features that are near one another have fewer similarities than features that are farther apart, the pattern shows negative spatial autocorrelation. The featureless plain of neoclassical economic theory is taking on new dimensions as scholars apply spatial statistics, with and without geographic visualization, to their research.

USING GIS FOR HISTORY

Many chapters in this book give examples of using digital overlay to bring together diverse sources of historical information about a place. Aaron C. Sheehan-Dean, for example, explains how the GIS built as part of the Valley of the Shadow project has enabled researchers to examine economic and political differences between two antebellum counties, using overlay and other kinds of spatial analysis to combine historical maps, manuscript census data, soil surveys, elevation data, precinct voting returns, and other information. In their chapter on historical maps, David Rumsey and Meredith Williams give special attention to the procedure called *rubber sheeting,* which digitally stretches and shrinks electronic versions of historical maps to make their features more nearly align with today's precise coordinate systems. Accurate registration of one layer atop another is essential for most kinds of spatial analysis and for combining historical data with the vast amount of modern geographical data now available in digital form.

Converting source material into spatial and attribute data takes up the lion's share of time that goes into creating a historical GIS. Although GIS can be used quickly to make simple maps, students and scholars thinking of using it as a foundation for historical research should be prepared to invest considerable time in designing their systems, acquiring data, and converting material from manuscript and print sources, including paper maps, into digital form. The rewards of such investment are not only being able to combine and jointly analyze diverse sources, but being able to map one's material in the course of research. Several of the essays here present the results of iterative mapping of large data sets. Andrew A. Beveridge and his fellow researchers made particularly intensive use of exploratory mapping in their study of the shifting patterns of ethnic and racial concentrations during the last one hundred years of metropolitan New York's tremendous geographical expansion. They and other contributors to this book found that using GIS as an investigative tool made them think more carefully about the significance and causes of geographical patterns, inspiring different questions than they might have asked otherwise.

GIS is particularly useful when the position of historic artifacts, buildings, roads, or other features is intrinsic to understanding their historical significance. Several essays in this volume demonstrate this point, including David W. Lowe's discussion of Civil War battlefield interpretation and Trevor M. Harris's overview of archaeological applications of GIS. If the study area and number of observations are fairly small, one can use a *Global Positioning System* (GPS) device to record locations in the field. If *ground truthing* is impractical or impossible, one must compare present-day maps, which provide accurate geographical coordinates for surviving landmarks, to historical maps that show the often only approximate location of features that have vanished from the landscape.

For some historical projects, producing maps is less important than compiling and organizing information from diverse sources according to geographical location. A simple map of a village and its outlying farms can be the first page of a virtual book about the place if one links locations on the map to historical photographs, letters from residents, extracts from local histories, cemetery and church records, and so on. Many applications of GIS to teaching history, like Benjamin C. Ray's Salem

Witch Trials Archive, exploit the technology's ability to integrate text, images, maps, and even sound in a single electronic space. The Electronic Cultural Atlas Initiative, discussed in Lewis R. Lancaster and David J. Bodenhamer's chapter, is a global project promoting the development of GIS-based digital archives and the dissemination of geographically referenced historical research.

The logical structure of GIS is ideally suited to systematic quantitative data, such as population censuses, land records, and social surveys. Some of the strongest contributions historical GIS projects are making to historiography are coming from quantitative historians using the technology to reexamine historical sources and to map previously unmapped evidence. Among this book's contributors, Amy Hillier reconsiders the redlining thesis by mapping Depression-era lending in Philadelphia. Geoff Cunfer finds that the patterns revealed by mapping meteorological data challenge the popular view of Dust Bowl history. In their essay on the Great Britain Historical GIS Project, Ian N. Gregory and Humphrey R. Southall show that taking account of historical changes to administrative boundaries will make population history more accurate and enable demographic historians to sharpen their understanding of migration. Alastair W. Pearson and Peter Collier show how researchers can refine their approach to land-use history by combining standard statistical techniques with a spatial statistical model that specifically addresses geographical variation. Like the cliometricians of a previous generation who used statistical methods and economic models to challenge historical understanding of American slavery, scholars at the forefront of historical GIS are finding that applying GIS to history can lead to new interpretations.

Historians are also discovering that incorporating time into GIS (sometimes called temporal GIS) can be quite difficult, particularly if one attempts to include time in spatial analysis. The simplest way to represent time in GIS is to enter relevant dates as attributes in the GIS database. One can then use date-based queries to map change in a sequence of images, like the frames of a movie. The Great Britain Historical GIS Project goes a step farther by capturing temporal changes to British administrative unit boundaries. Each boundary line is assigned a start date and an end date. The GIS can then generate maps that accurately show all boundaries as they stood at a particular date, say January 1, 1881, or June 14, 1895. Things become more complicated if one wants to map changes that occurred across two or more GIS layers, or changes that affected some objects within the database but not others. Historians may claim a special interest in understanding time, but their wish for better temporal analysis is shared by physical scientists who would like GIS to map the varying damage caused by mud slides after major storms, and urban planners who would like better models to predict changes in land use and rainwater runoff in the wake of a new highway construction project. We all want GIS to simulate the complexity of real-world environments and analyze how and why they change when they change. A tall order for any technology.

Computer database programs in general do not readily accommodate the ambiguity, incompleteness, and bias—in short, much of the social meaning—of historical sources. Quantitative historians have long struggled with these limitations. In GIS, the requirement of locational precision raises further problems. Can a historical GIS preserve the imprecision of location that characterized geographical knowledge before the age of satellite imagery? Can one represent the vagueness of historical sources and so prevent the illusion of certainty in the maps, charts, and tables produced from a GIS? Some scholars working in historical GIS are pursuing answers to these questions, and in geographic information science the concepts of uncertainty and imprecision are becoming increasingly important. Tom Elliott and Richard Talbert explain how the Ancient World Mapping Center is grappling with the fundamental issue of how to represent the uncertainty of location and time in scholarship on Greek and Roman civilization.

The essays here suggest that studies based on geographical analysis differ from conventional history in a number of ways. They tend to emphasize context over chronology and to bring more elements into the frame. They often have a strong local or regional focus. Their attention to the interactions of many factors within a given space or place is akin to the perspectives of environmental history, regional history, and historical geography. Historical GIS often explores relationships between localities and between one scale of human interaction and another, as between local, regional, and national conditions and events. Sensitivity to the scale of analysis helps historical scholars clarify the level at which certain explanations work well and where they become less convincing. As more historians adopt GIS, the care they take in contextualizing and exposing the biases of historical sources will more deeply inform geographical scholarship. And geographical knowledge and skills will begin to play a role in historical training.

Contributors to this book come from a wide variety of academic backgrounds. The nine geographers, including myself, were familiar by training and experience with mapping social phenomena, but for some the work here represents early, experimental attempts to apply GIS to historical questions. For the other contributors, who come from history, religious studies, sociology, social work, and the history of cartography, GIS has been a more radical departure. Authors' use of GIS has ranged from solo projects looking at particular topics to team efforts of many years' duration aimed at constructing digital archives or other kinds of digital infrastructure that will support future scholarship. For everyone, the decision to use GIS has raised important problems and exciting possibilities. We have all been frustrated and exhilarated while mastering new skills, have all faced the decision of whether and how to collaborate, and have all struggled to reconcile historical sources with a twenty-first-century technology. We hope readers will join us in taking on these challenges and discovering new ones.

Acknowledgments

For their very helpful reviews of manuscript chapters, I would like to thank Brian Q. Cannon, Michael Curry, James A. Henretta, Roger J. P. Kain, Peter Knupfer, Philip Kohl, John R. Langton, Mary Lefkowitz, Robert D. Lewis, and John Long. Special thanks to Michael F. Goodchild and Gary Kornblith for their careful reviews of the entire manuscript. I would also like to thank Wellesley College for funding my own early education in GIS. At ESRI Press, thanks to Michael Hyatt for his excellent design and to Christian Harder for his enthusiastic support of this project. Michael Karman's implacably high standards, perceptive editing, and collegial spirit made it a joy.

CHAPTER ONE

HISTORICAL MAPS IN GIS

David Rumsey and Meredith Williams

Most historical GIS would be impossible without historical maps, as the chapters in this book testify. Maps record the geographical information that is fundamental to reconstructing past places, whether town, region, or nation. Historical maps often hold information retained by no other written source, such as place-names, boundaries, and physical features that have been modified or erased by modern development. Historical maps capture the attitudes of those who made them and represent worldviews of their time. A map's degree of accuracy tells us much about the state of technology and scientific understanding at the time of its creation. By incorporating information from historical maps, scholars doing historical GIS are stimulating new interest in these rich sources that have much to offer historical scholarship and teaching. At the same time, the maps themselves challenge GIS users to understand the geographic principles of cartography, particularly scale and projection. We have addressed these challenges in order to examine the value of including nineteenth- and early twentieth-century paper maps in GIS.[1]

One can use digital renditions of historical maps to study historical landscapes, the

Figure 1. Wheeler Survey map of Yosemite Valley, 1883
The government-funded Wheeler Survey produced one of the first accurate maps of Yosemite Valley. The cartographers who drew the map used hachuring (a form of shading) to suggest the depth of the canyon and the river valleys leading to it.

maps themselves, and how places changed over time. GIS is breathing new life into historical maps by freeing them from the static confines of their original print form. It is also enabling a new level of understanding. Traditionally, people read and analyzed maps using a critical eye and *a priori* knowledge. Comparison of two or more maps was possible, but the conclusions were only as reliable as the reader's visual acuity and interpretive skill. The same limits applied to cartography, the making of maps. Cartographers traditionally made maps by gathering information from published maps or field surveys. The maps they produced were often marvelous acts of interpretation. Consider the Wheeler Survey, which mapped territory west of

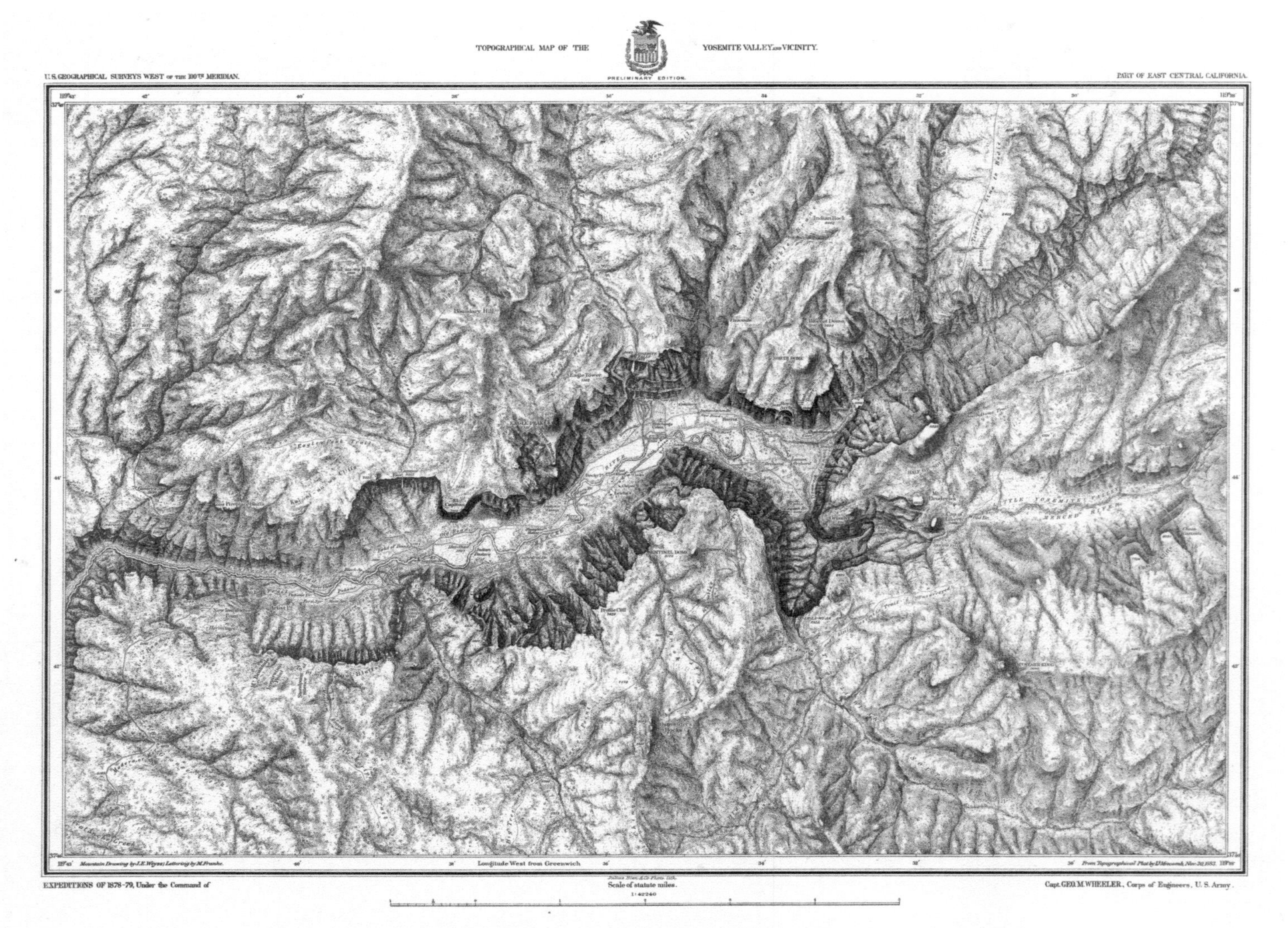

the 100th meridian in a series of geographical expeditions beginning in 1871. George M. Wheeler and his associates carried heavy plane tables, surveyor's instruments, and large-format cameras by wagon and mule across mountains, canyons, and deserts. Their survey points were highly accurate for the time, but the renderings of the topography that linked those points in a continuous landscape were as much art as science *(figure 1)*. When their map is converted into digital form, it can be manipulated and combined with other spatial data, such as digital elevation models *(figure 2)*. The three-dimensional landscape is more immediately recognizable. It gives us the feel of standing next to the cartographer as he gazed over Yosemite Valley.

Figure 2. Wheeler's Yosemite Valley in 3-D

Draping the scanned image of the original Yosemite map over modern digital elevation model maps gives the old map a new look and immediacy. The simulated depth of the 3-D terrain model complements the beautiful hachuring of the 1883 map. In this map we used a vertical exaggeration factor of 1.5.

Figure 3. Chicago in 1868 and 1997
Hot links connecting historical city plans to present-day maps give students easy access to visual comparisons. Rufus Blanchard's Guide Map of Chicago (1868) shows the city's characteristic grid and cross-cutting diagonal arteries, little changed in a recent ArcView® StreetMap™. The dimensions of the city, however, have changed enormously.

More importantly for the aims of historical research, information that was difficult to perceive in the historical map is now accessible for our own investigation. We can now measure elevation, distance, and area, and rotate the image to place ourselves at different viewpoints.

Ordinarily, the first step in preparing a paper map for use in GIS is scanning it. For this purpose, it is best to capture map images at a very high resolution.[2] If one's main purpose is to study maps as historical documents, scanning may be all the manipulation required. Scanned maps can be easily incorporated into a GIS as graphic images. Connected by hot links to particular features in a GIS layer, historical maps can be opened to compare present and past configurations of a given place or landscape *(figure 3)*.

Integrating historical maps in GIS to analyze the spatial information they contain, or to layer them with other spatial data, requires that the maps be georeferenced. That is, selected *control points* on a scan of the original map must be aligned with their actual geographical location, either by assigning geographical coordinates to each point, or by linking each point to its equivalent on a modern,

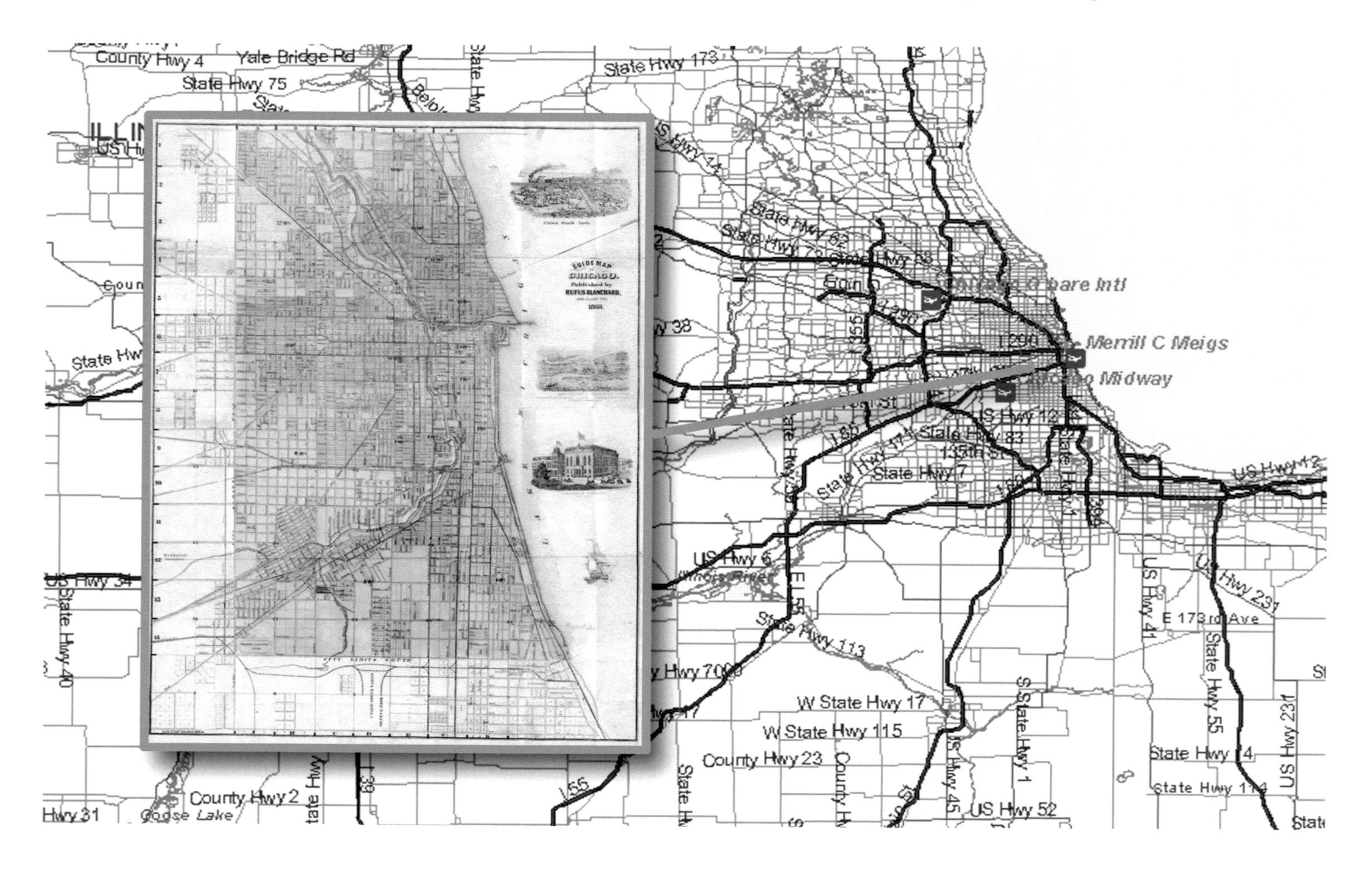

accurate digital map. Once the control points are in place, one applies mathematical algorithms to warp the original map image to fit the chosen map projection as nearly as possible. Further adjustments can be done manually to try to find the best fit for all parts of the original map. The process is sometimes called *rubber sheeting* because it stretches and shrinks the map image like a thin sheet of rubber being pulled to fit a particular form. Figure 4, for example, is a scanned detail of the Wheeler Survey's 1882 map of New Mexico that we georeferenced and then layered with point data marking the location of Jesuit missions.[3] Placing the missions in their physical context helps one see the geographical logic of their location and infer their relationship with regional patterns of commerce and transportation that developed between their founding in the late seventeenth century and their decline at the end of the nineteenth.

It is almost impossible to perfectly align an old map to modern coordinate systems because mapping methods before the age of aerial photography often only very imprecisely represented scale, angle, distance, and direction. For most GIS projects, the value of the historical information

Figure 4. Adding features to a historical map

This 1882 georeferenced map from the Wheeler Survey shows the Rio Grande Valley just north of Santa Fe. The cartographer's hachures mark the steep plateaus and mesas of the rugged region where Franciscan priests established missions among the native Indian people as early as 1692. The red points represent a layer of database information about the missions. By clicking on the points, users can bring up historical data about each site.

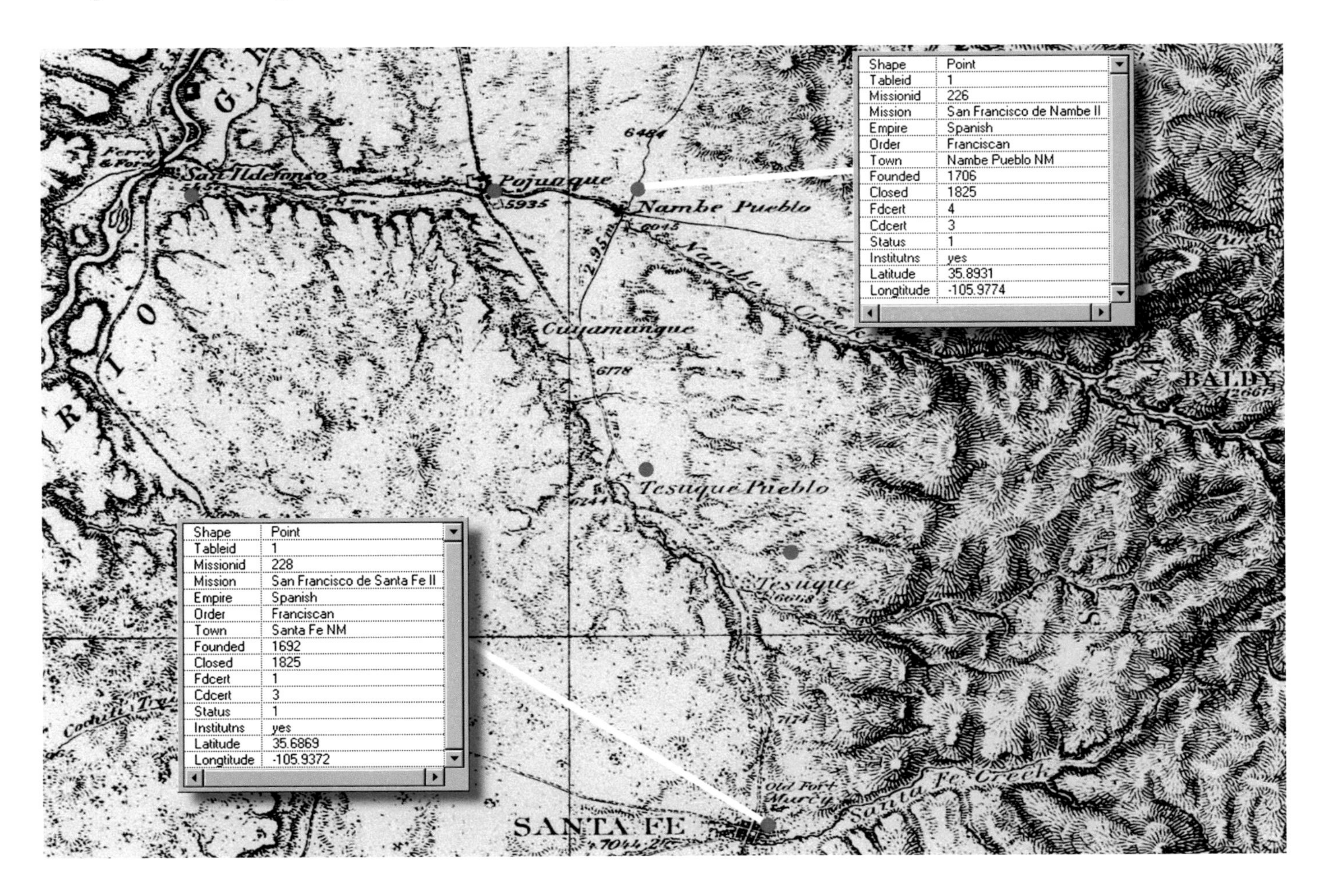

on paper maps more than compensates for the residual error in their georeferenced versions. What one should keep in mind is that georeferencing does not necessarily improve a historical map or make it more accurate. In the course of changing the original map to make it amenable to digital integration, georeferencing changes lines and shapes, the distance between objects, the map's aesthetics, and its value as a cultural artifact. One gains knowledge of the original while processing it for inclusion in GIS, but one also loses something if the original map is not represented for comparison with its actual size, proportions, and qualities. Ideally, researchers should include both the warped map and the scanned image of the original map in a GIS project or publication.

The impossibility of aligning historical maps perfectly with modern maps can itself yield historical information. During their exploration of the Louisiana Purchase, Meriwether Lewis and William Clark made notations of longitude and latitude, which they later used in drawing the monumental map of their expedition *(figure 5)*. To bring this map into GIS, we began by estimating the projection used for Lewis and Clark's rendering. Their map appears to be on a conic projection, since the lines of latitude are curved and those for longitude converge. A common conic projection for maps of the continental United States is Albers Equal-Area, so we used a vector layer in this projection as a base for georeferencing the exploration map. Control points were taken by assigning geographical coordinates to reference points, such as the explorer's longitude and latitude marks and major river junctions. By then overlaying modern highways (red) and state boundaries (brown), one can gauge Lewis and Clark's errors in measurement and estimation.

Small-scale historical maps like Lewis and Clark's are prone to greater error than large-scale maps.[4] A large area of the earth's surface, like the continental United States, is harder to depict on a flat surface than a small area such as a town.[5] Large-scale maps are often more easily and accurately converted for use in GIS because they tend to have less egregious geographical errors. Maps drawn at a large scale may also have more unique local data to contribute, such as information on land ownership, the location of buildings, paths, and streams, and so forth.

Large-scale city maps can be wonderful sources for urban history, and adding one to a GIS can greatly increase its utility. Many American cities were mapped repeatedly and in great detail in the nineteenth

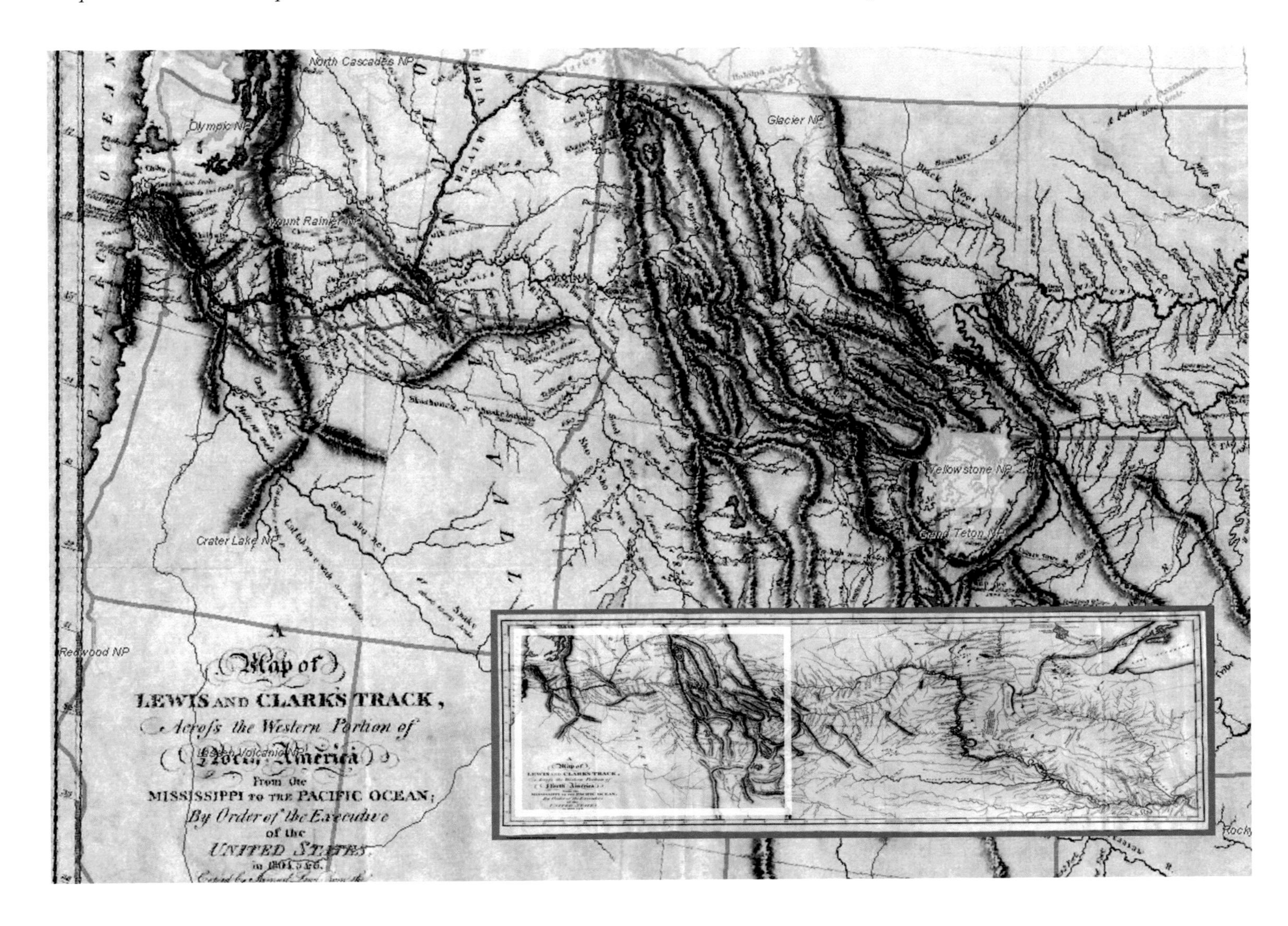

Figure 5. Gauging the accuracy of Lewis and Clark's map
A Map of Lewis and Clark's Track, Across the Western Portion of North America (1814, inset, lower right) combines the explorers' observations with reports of rivers, mountains, and other features from Native Americans and other people Lewis and Clark encountered on their three-year journey. The original map is 70 cm wide. Georeferencing the map reveals significant distortion in the position of the western coast (note the brown outline of the true western coast on the left side of the figure). This was likely due to errors in drafting or projection of the original map.

and twentieth centuries, none more so than New York. J. H. Colton's *Topographical Map of the City and County of New-York* (1836) and Matthew Dripps' *Map of the City of New York* (1852) provide unparalleled images of the built environment of America's leading nineteenth-century commercial city. These two unusually large maps are of differing scales. Colton's map, at a scale of 1:15,840, measures 29 by almost 70 inches. Dripps' huge map, nearly 88 inches long and 46 inches wide, shows lower Manhattan at the very large scale of 1:3,450, approximately two inches to the city block. Although the physical size of the two maps made them difficult to scan, it was possible, and their large scales and precision made them relatively easy to georeference. Overlaying georeferenced historical maps allows one to combine maps of greatly differing sizes and scales, such as these, in the same coordinate space.

With the innovation of partially transparent raster layers in GIS, we were able to produce composite map images that suggest how the passage of time transformed the city *(figure 6)*. Some changes became strikingly clear when we then overlaid a modern digital street map on the georeferenced Colton map *(figure 7)*. For example, we could see that the Hudson Parkway and East River Drive, highways that run along the west and east sides of Manhattan, respectively, were built on landfill beginning at about 72d Street. The original New York City reservoir, whose ghostly rectangular pools occupy the center of Central Park on the map detail, was moved north and given a more natural-seeming shape by the park's landscape architect, Frederick Law Olmsted. What was low-lying, swampy ground in the early 1850s, on the upper East Side between 85th and 105th streets, was drained and filled later in the century to provide housing for the city's exploding population.

Visual overlay of this type is very useful in research and teaching. However, to query or measure spatial relationships between features, they must be lifted off historical maps and made into vector GIS layers. This is done by digitizing map features as points, lines, and areas. Because few archives provide access to a digitizing tablet, scholars often choose to have historical maps scanned. They can then digitize the maps directly on screen. Digitizing is far more time-consuming than georeferencing, but it adds tremendously to the amount of data available for use in GIS. By creating vector polygon features from the city blocks and building locations on Colton's New York City map,

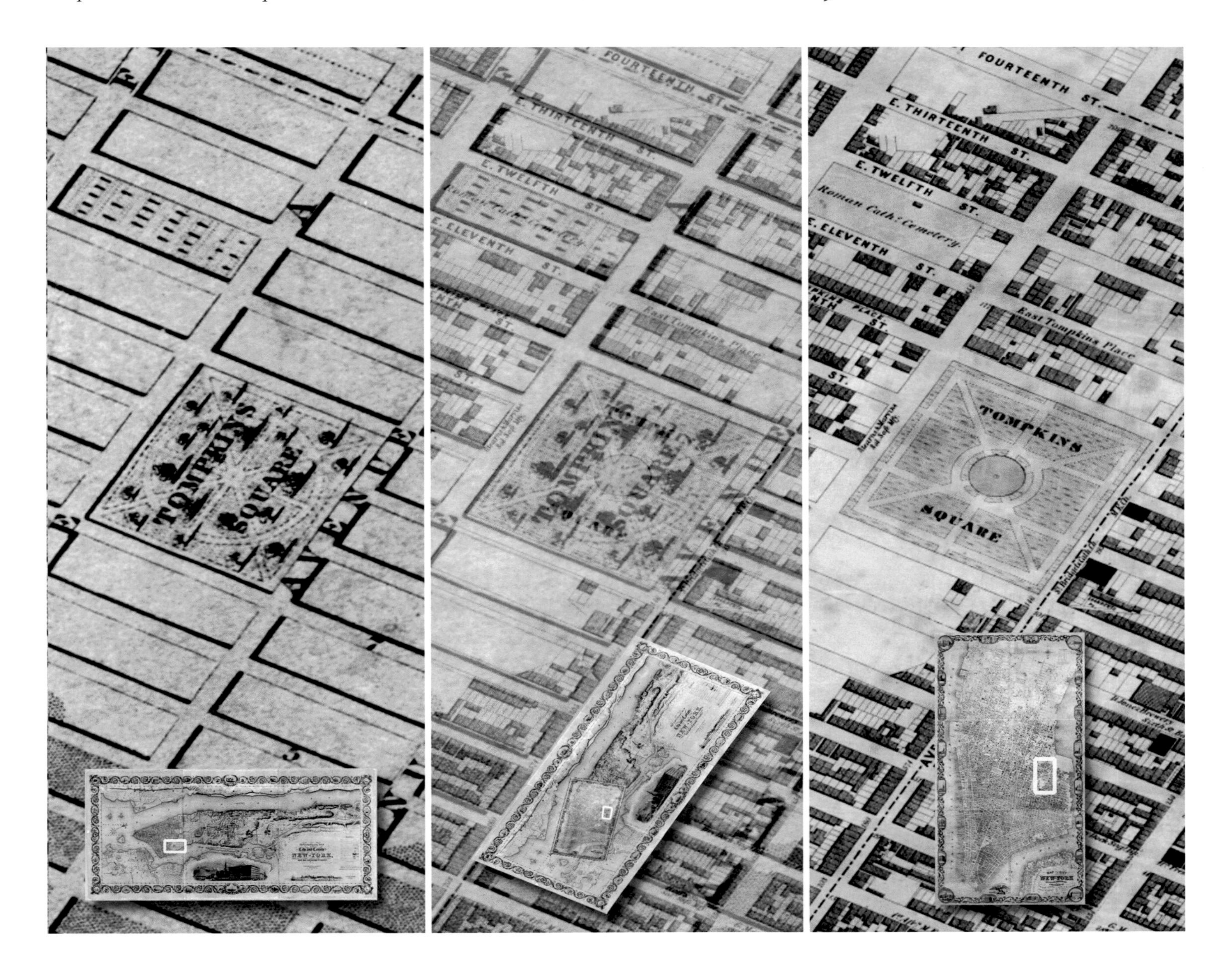

Figure 6. Urban development in New York City, 1836–1852

In 1836, the area around Tompkins Square in the East Village had few occupants but those buried in the Roman Catholic Cemetery, as seen in the map on the left. By 1852 (right map) the neighborhood was full of row houses, businesses, tenements, and back alleys where servants and low-wage workers lived. Tompkins Square had also been transformed from a country park of trees and ornamental paths to an open green with a central fountain. A transparent composite of the two maps (center image) makes comparison easier.

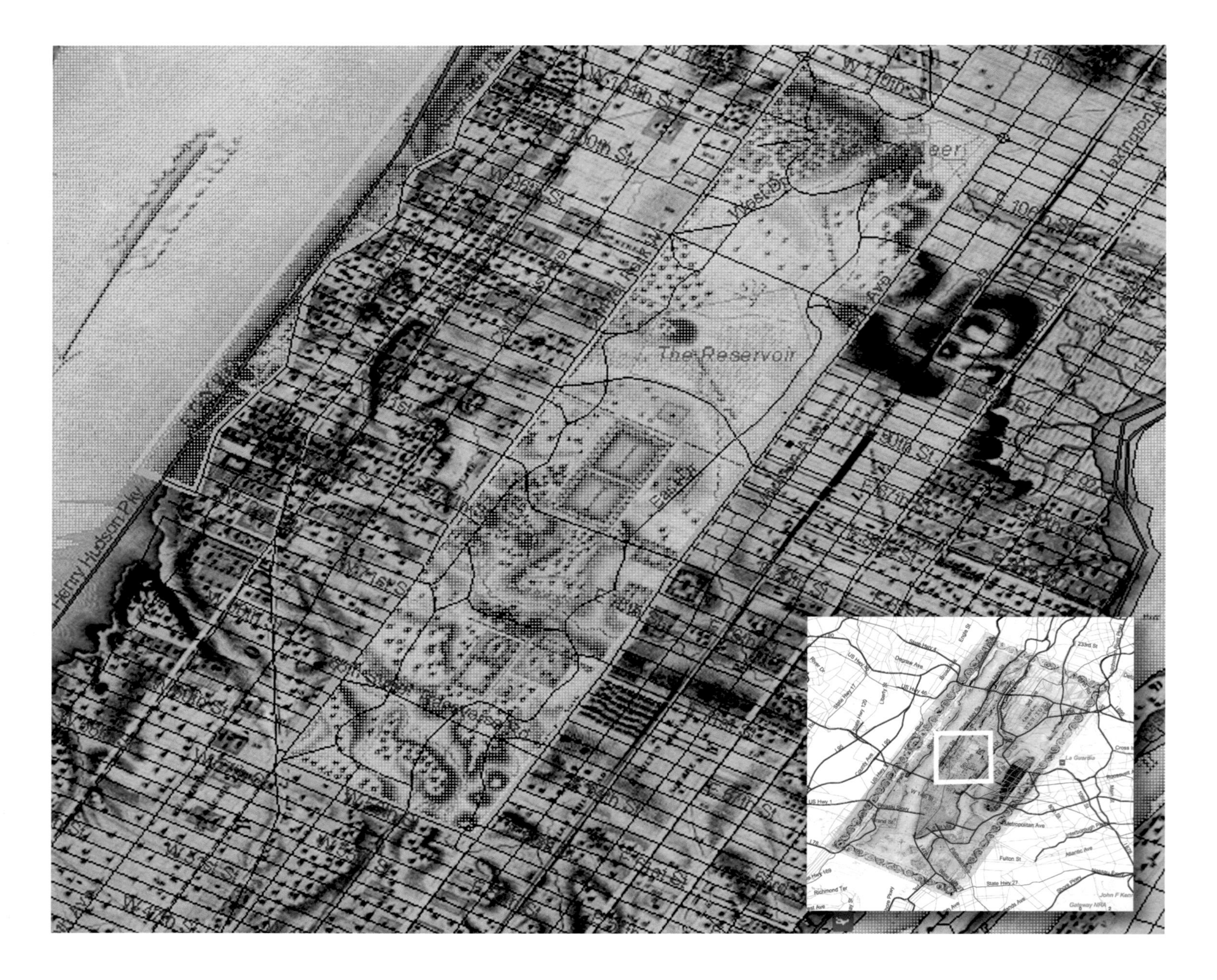

Figure 7. Comparing New York City in 1836 and 1997
GIS enables one to compare georeferenced historical maps with modern maps by overlaying them. This image overlays a modern street map on Colton's map of the antebellum city. All of Manhattan was platted for development in 1836, but above Times Square the grid had more trees than people.

for example, we could join attribute data such as owner, land-use category, date of construction, structure type, and architectural style for each building. All lots could then be queried for ownership and classified as either private or public. Different colors could then be used to signify public and private lots. Total acreage could be calculated for each land category. Performing the same steps on a later map of the city and comparing the two sets of data would enable us to identify the changing patterns of public and private ownership and architectural style. The dynamics of urban change could be displayed through an animation of cartographic snapshots of New York's built environment.

In addition to illuminating urban history, historical maps can provide valuable information about environmental change. Today, satellites circle the earth recording daily changes on the planet's surface. Scientists study satellites' remotely sensed data to determine how human activity and natural phenomena interact. Historical maps make it possible to extend the examination of humans' environmental impact far back before the advent of satellite imagery. To explore changes in land use in southern California, we used GIS to combine the 1881 Wheeler Survey thematic map of the area with a 1998 land-cover map from the California Gap Analysis Project *(figure 8).* The Wheeler map depicts a rural landscape scarcely touched by urbanization. It displays five land-use categories: agriculture, timber, grazing, arid/barren land, and chaparral, shown in pastel tints of yellow, green, and gray. The city of Los Angeles, with just eleven thousand people, sits at the base of the Santa Monica and San Gabriel mountains. We overlaid the 1998 data as a transparent polygon layer to examine the extent of the land-use changes. Wheeler's team would be surprised to see that today, urban development covers all but the highest, steepest peaks.

On a number of the historical maps we have discussed, shading or hachures (fine black lines) suggested elevation. Using GIS, one can simulate topography more dramatically and vividly by using *digital elevation models,* which are raster surfaces composed of longitude (x), latitude (y), and elevation (z) coordinates.[6] We already saw how draping the Wheeler Survey's Yosemite map over a digital elevation model enhanced the historical map's depiction of the landscape. Verisimilitude can be a powerful teaching tool when it helps students understand the physical character of a past place. When we displayed August Chevalier's 1915 map of San Francisco over

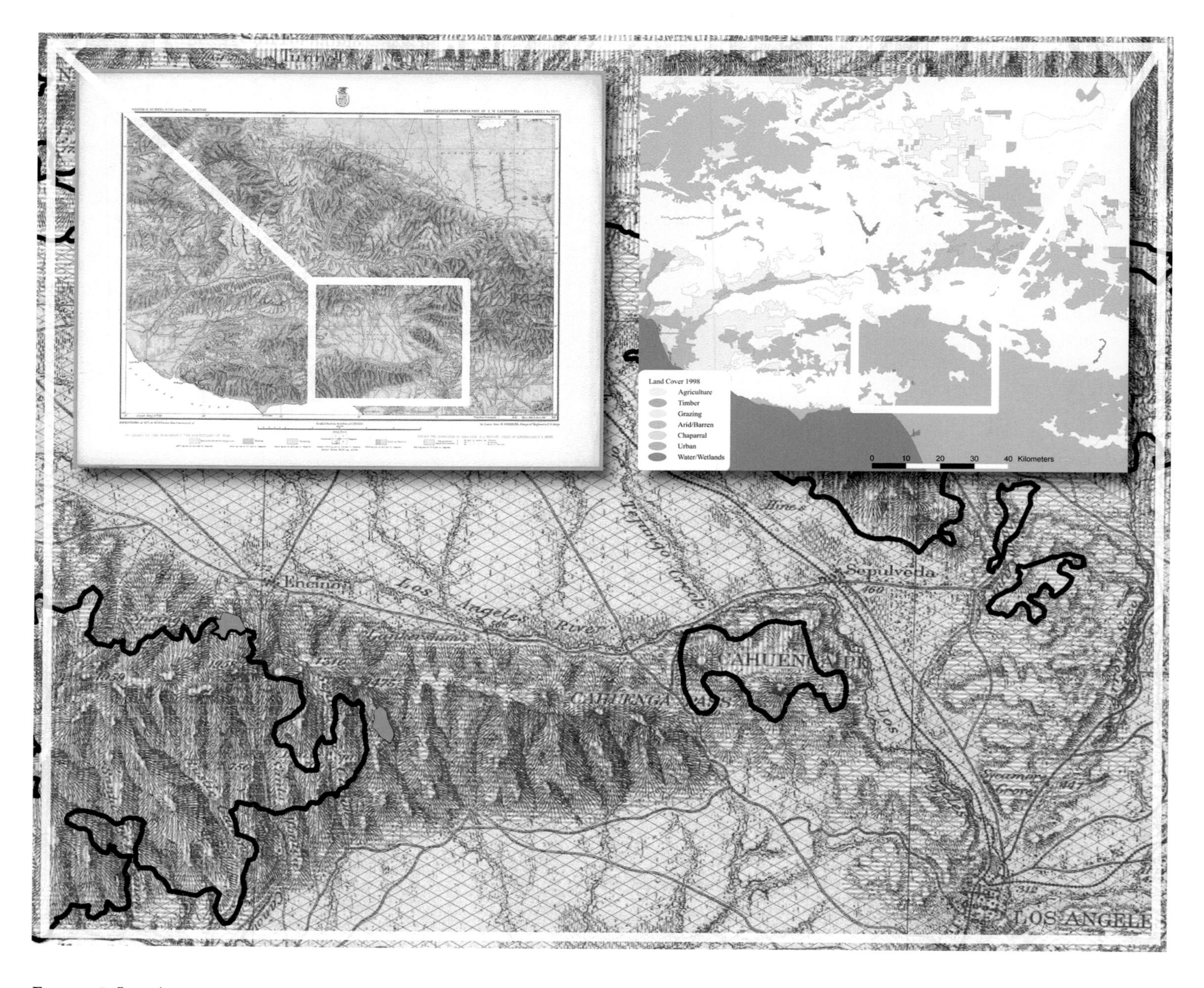

FIGURE 8. LOS ANGELES THEN AND NOW

LOOKING BACK AND FORTH BETWEEN THE TWO UPPER MAPS GIVES ONE A SENSE OF HOW MUCH THE SOUTHERN CALIFORNIA LANDSCAPE HAS CHANGED. THE WHEELER SURVEY MAP OF 1881 (UPPER LEFT) SHOWS A REGION DOMINATED BY AGRICULTURE (YELLOW), GRAZING (LIGHT GREEN), TIMBER (DARK GREEN), AND ARID LAND (LIGHT BROWN). THE 1998 MAP (UPPER RIGHT) SHOWS A FLOOD OF URBAN DEVELOPMENT (GRAY) COVERING MOST OF THE FORTY-KILOMETER AREA OF THE LOS ANGELES BASIN, OUTLINED IN WHITE. IN THE COMBINED MAP SHOWN AT A LARGER SCALE, URBAN DEVELOPMENT, SYMBOLIZED BY A RED DIAGONAL GRID, COVERS ALMOST ALL OF THE BASIN.

a digital elevation model, the extreme topography of the city leapt to view *(figure 9)*.

One can also use digital models to explore the terrain that lies beneath the oceans. We experimented with this idea using our digital rendering of the 1926 U.S. Coast Survey map of San Francisco Bay and the 1998 digital contour line map of the same area produced by the National Oceanic and Atmospheric Administration (NOAA). The 1926 map noted depth at hundreds of points around the bay. When we queried the same points on the 1998 map, we found that some elevations had changed significantly due to human activity. The north end of Treasure Island, for example, which lay beneath fourteen fathoms of seawater in 1926, was now above sea level; landfill more than tripled the island's dry surface area. To clearly display the *bathymetry* of the bay—that is, the shape and depth of the seafloor—we converted the NOAA contour map into a *triangulated irregular network* (TIN) layer.[7] This created a detailed 3-D model of the bathymetry that we could then combine with the historical Coast Survey map *(figures 10, 11)*. Suddenly we could easily and vividly see the deep trench of

Figure 9. San Francisco in 3-D
August Chevalier used shading to suggest the rugged topography of San Francisco in 1915. Draping a georeferenced version of his map over a digital elevation model more clearly shows the relative height of each hill, the valleys cut by streams, and the density of business and residential development on the most level ground along the harbor. The vertical exaggeration factor in this map is 1.5.

Figure 10. The depths of San Francisco Bay

Adding visual depth to the bay puts two historical icons of San Francisco into context: the Golden Gate Bridge, its claim to be an engineering marvel now evident in the four-hundred-foot chasm it spans; and the top-security prison on Alcatraz Island, guarded by deep waters and swift currents. Because the 3-D model was built from bathymetry data only, the dry land appears flat. We used a vertical exaggeration factor of 7 to bring out depth variation in this map and figure 11.

Figure 11. San Francisco Bay showing Treasure Island landfill

The 3-d model reveals how landfill, apparent in the TIN underlying surface, greatly expanded the north end of Treasure Island for the 1939 World's Fair. In the same area, markings from the original sea chart, drawn thirteen years earlier, show the prefill depths of fourteen to twenty-two fathoms.

the Golden Gate Strait, the relative shallows of San Pablo Bay, and the chasm surrounding Alcatraz Island, site of the famous prison. The 3-D viewing software even allowed us to travel around the scene as if in a submarine.

To include historical maps in GIS for teaching and research, scholars will need access to high-quality scanners or prepared digital images. We envisage a time in the near future when thousands of historical map images, some already georeferenced and digitized as vector layers, will be available through shared networks and public-access Web sites. Recent improvements in GIS software are resolving the problems of storing ever-larger spatial data sets by enabling users to access remotely stored data. The Library of Congress Geography and Map Division and a few other leading map collections in the United States and other countries have launched digital dissemination projects. From the Library of Congress Memory Web site, anyone can download map images and then zoom in to study their details on-screen as if with a powerful magnifying glass. The David Rumsey Historical Map Collection has scanned more than sixty-five hundred historical maps and made them available online. Rumsey will provide an increasing number of georeferenced versions in the next few years. Digital renderings of historical maps are also available through ESRI's Geography Network and the Electronic Cultural Atlas Initiative's Metadata Clearinghouse. Since the process of converting historical maps into GIS-compatible formats is time-consuming, resource intensive, and expensive, it is doubly important that the burden be shared and the resulting resources aggregated. Each digitized map will require excellent, standardized metadata to describe it and make it easy to retrieve.

Historical maps have a great deal to offer GIS, and GIS brings new techniques to the analysis and display of historical maps. As historical maps become more widely available and as GIS becomes more sophisticated, it is certain that scholars will combine the two in creative ways yet to be imagined. Cartographers of days past would have been pleased to know that centuries later a new mapping technology is stimulating new interest in their work.

Map sources

All maps courtesy of the David Rumsey Historical Map Collection, www.davidrumsey.com.

Blanchard, Rufus, *Guide Map of Chicago,* 1868. Chicago: Rufus Blanchard. Size: 53 cm × 41 cm. Scale: 1:25,344.

Colton, J. H., *Topographical Map of the City and County of New-York,* 1836. New York: J. H. Colton & Co. Size: 74 cm × 170 cm. Scale: 1:15,840.

Chevalier, August, *The "Chevalier" Commercial, Pictorial and Tourist Map of San Francisco,* 1915. San Francisco: Aug. Chevalier. Size: 160 cm × 145 cm. Scale: 1:9,600.

Dripps, Matthew, *Map of the City of New York Extending Northward to Fiftieth St.,* 1852. New York: M. Dripps. Size: 223 cm × 117 cm. Scale: 1:3,450.

Lewis, Meriwether, and William Clark, *A Map of Lewis and Clark's Track, Across the Western Portion of North America,* 1814. Philadelphia: Bradford and Inskeep. Size: 30 cm × 70 cm. Scale: 1:4,350,000.

United States Coast and Geodetic Survey, *United States–West Coast. San Francisco Entrance, California,* 1926. Washington, D.C.: U.S. Coast and Geodetic Survey. Size: 85 cm × 105 cm. Scale: 1:40,000.

Wheeler, G. M, *Land Classification Map of Part of S. W. California, Atlas Sheet No. 73 (C),* 1881. Washington, D.C.: U.S. Government. Size: 45 cm × 52 cm. Scale: 1:253,440.

Wheeler, G. M., *North Central New Mexico, Atlas Sheet No. 69 (D),* 1876. Washington, D.C.: U.S. Government. Size: 45 cm × 51 cm. Scale: 1:253,440.

Wheeler, G. M., *Topographical Map of the Yosemite Valley and Vicinity,* 1883. Washington, D.C.: U.S. Government. Size: 43 cm × 55 cm. Scale: 1:42,240.

Software and hardware

ArcView 3.2, ArcView 8.1, Adobe® Photoshop® 5.5, Luna Imaging, Inc.'s Insight®, MrSID™ image compression software by LizardTech®, PhaseOne image-capture software. PhaseOne PowerPhase and PowerPhase FX 4×5 digital scanning camera back, Sinar X 4×5 view cameras with Rodenstock lenses, Kaiser RePro copy stand with Videssence Fluorescent Icelites, Apple® Macintosh® G4 and G3, Gateway™ workstation, DVD storage disks.

Further reading

Brown, Lloyd A. *The Story of Maps.* New York: Dover, 1977; orig. pub. 1949.

Harley, J. B. *The New Nature of Map: Essays in the History of Cartography.* Paul Laxton, ed. Baltimore: Johns Hopkins University Press, 2001.

Paullin, Charles O. *Atlas of the Historical Geography of the United States.* John K. Wright, ed. Carnegie Institution of Washington Publication no. 401. Washington, D.C.: Carnegie Institution of Washington and American Geographical Society of New York, 1932.

Raisz, Erwin. *General Cartography.* 2d ed. New York: McGraw-Hill, 1948.

Continued

Further reading (continued)

Reps, John William. *Views and Viewmakers of Urban America: Lithographs of Towns and Cities in the United States and Canada, 1834–1926.* Columbia, Mo.: University of Missouri Press, 1984.

Robinson, Arthur H. *Early Thematic Mapping in the History of Cartography.* Chicago: University of Chicago Press, 1982.

Short, John R. *Representing the Republic: Mapping the United States 1600–1900.* London: Reaktion Books, 2001.

Snyder, John P. *Flattening the Earth: Two Thousand Years of Map Projections.* Chicago: University of Chicago Press, 1993.

Woodward, David, and others, eds. *History of Cartography,* multiple volumes. Chicago: University of Chicago Press, beginning 1987.

Online resources

David Rumsey Historical Map Collection: www.davidrumsey.com

Electronic Cultural Atlas Initiative Metadata Clearinghouse: ecai.org/tech/mdch.html

ESRI Geography Network: www.geographynetwork.com

Library of Congress Memory Web site: memory.loc.gov/ammem/gmdhtml/gmdhome.html

---

Notes

1. Over the course of history, maps have been produced in many other media: static and mobile; written and oral; in two, three, and four dimensions.
2. To hold the detail in historical maps, 600 pixels per inch are often required. Because high-resolution scans result in very large raster files, frequently over one gigabyte, image compression is required to ease the transmission of files and their incorporation into GIS.
3. John Corrigan and Tracy Neal Leavelle, *Catholic Missions in Colonial North America* (Berkeley, Calif.: California Digital Library for the Electronic Cultural Atlas Initiative, forthcoming).
4. Map scale is the ratio of a distance on a map to the corresponding distance on the ground. The larger the ratio, the smaller the scale. A small-scale map of the world might have a ratio of 1:5,000,000, while a large-scale city map might have a ratio of 1:25,000.
5. Daniel Dorling and David Fairbairn, *Mapping: Ways of Representing the World* (Dorchester, Dorset: Addison Wesley Longman Limited, 1997), 28–38.
6. In order to best represent topography, we used a vertical exaggeration factor of 1.5 when displaying Chevalier's San Francisco map and Wheeler's Yosemite Valley map in 3-D. The 3-D bathymetry map of San Francisco was created using a vertical exaggeration factor of 7. Since this exaggeration is applied equally to the entire surface, the relationship between depths or heights in a particular landscape remain proportional.
7. In a TIN, a form of the tesseral model, the vertices of the non-overlapping triangles used to represent a surface are irregularly spaced nodes. Each node has an x,y coordinate and a surface, or z, value. Unlike a grid, a TIN allows dense information in complex areas and sparse information in simpler or more homogeneous areas, a characteristic that makes it suitable for modeling highly variable surfaces such as the chasms and plains of the seafloor.

CHAPTER TWO

TEACHING THE SALEM WITCH TRIALS

Benjamin C. Ray

EARLY in 1692, a handful of girls put a small village in Massachusetts under a spell that would last well into the next year and would engulf a good portion of eastern Massachusetts Bay Colony. As people in other towns joined in, judges in the colony heard accusations against at least 168 people: young and old, men and women, ministers and merchants, leaders and derelicts. By October of 1692, nineteen had been found guilty of witchcraft and hanged, one was pressed to death with stones, and five died in prison. No one who confessed to witchcraft was put to death. Later that month, Governor William Phips shut down the witchcraft court, and from January through May 1693, the Supreme Court began to clear jails of the accused. There were no more convictions. The spell was over.

Few episodes in American history have gripped the imagination as powerfully as the Salem witch trials, from the impassioned pamphlet against witchcraft by Reverend Cotton Mather to equally strong protests against the trials by Thomas Brattle and Robert Calef; from Tompkins Matteson's dramatic painting of *The Trial of George Jacobs, Sr.,* in 1855 *(figure 1)* to Arthur Miller's drama, *The Crucible,* in 1953.

266

The deposition of Ann putnam who testifieth and saith
that I haue ben most greviously afflected by George Jacobs sen
but most dreadfully tormented by him on: 5t: of may 1692
dureing the time of his examination also on the day of his examination
I saw George Jacobs or his Appearance most greviously torment mary
walcot Eliz: Hubburd and I beleue in my hart that
George Jacobs is a dreadfull wizzard and that he hath very often
afflected me and the afore mentioned persons by his acts of wichcraft
ann putnam owned this har testimony before the Juriars of
Inquest: on har oath this 4 day of agust: 1692

Jurat in Curia

Figure 1. "The Trial of George Jacobs, Sr." and Ann Putnam's accusation
Tompkins Matteson used color, motion, and stillness to convey the passionate emotions of Salem's witch trials. George Jacobs, whom village teenager Ann Putnam accused of being a "dreadful wizzard," kneels before the stern, black-robed judges. Around them swirl angry and terrified Puritans shouting for justice. "Afflicted" Ann Putnam may be the girl at the bottom of the picture, pointing at Jacobs. Below the painting is a digital scan of her deposition, in which she claims to have been "most dreadfully tormented" by her old neighbor. Matteson painting courtesy the Peabody and Essex Museum; Putnam deposition courtesy the Massachusetts Historical Society.

So horrifying were the events and so unaccountable did the trials seem to later generations that writers and scholars from 1692 to the present have looked for ways to explain them, scouring court documents and other sources to find clues to the causes of the tragedy, studying, interpreting, and reinterpreting the evidence. When I set about to teach a seminar on the witch trials, I wanted to put the original documents in students' hands so that they, like professional historians, could experience the excitement of discovering history in contemporaries' own words. I thought, too, that students would understand the events in Salem better if they studied the original court records alongside scholars' accounts.

Because all collections of the primary sources were either out of print or otherwise unavailable, my plan required creating a digital library. A small teaching grant enabled me to digitize transcriptions of the court documents, as well as several pamphlets and books that appeared immediately after the trials. These documents were the first items in the Salem Witch Trials Archive,[1] which now contains approximately 850 legal documents and other primary source materials. I then decided to expand the electronic archive with funding from the

National Endowment for the Humanities and other sources, so that students and other researchers could use it to explore the witch trials.

One of the organizations from which I obtained support, the Electronic Cultural Atlas Initiative, required that all material in the archive be referenced geographically and chronologically. Using a geographic information system, we developed a database structure linking historical maps and other visual documents to demographic, genealogical, and legal material. Pulling together so many historical documents and types of information created a complex set of categories and relationships. What makes the archive coherent and useful is its simultaneously geographical and historical organization—the linking of every document, every image, and every piece of demographic and genealogical information to every person involved, with their location in place and in time.

I had no prior experience with relational databases or with GIS. As often happens, learning to use new tools pushed me to rethink assumptions about how to organize and use material in teaching and research. Since the technology I was using can manage vast amounts of material, it encouraged me to take an encyclopedic approach. If, for example, the names of all the accusers and the accused can be located on a map, that possibility invites the linking of other information to each name, such as dates of birth and death, family history, and economic information. Using the map, we could then show the pattern of the accusations as they happened over time and the links between accusers and accused. This raised questions about the events that have not been fully studied before. Most research has concentrated on Salem village. Mapmaking led me to wonder how much geographic territory should be covered in a study of the Salem witch trials. I also began thinking about what kinds of questions students would be able to pursue if they had access to geographical information about the witch trials, such as where those involved in the trials lived, how far they were from courts and jails, to which churches they belonged, and what proximity or distance might suggest about the relationships between accusers and accused.

I was by no means the first to recognize the importance of geography for understanding events in Salem. Maps have been used by scholars of the witch trials since Charles Upham, a local Salem historian, published a landmark two-volume history in 1867 titled *Salem Witchcraft.* Upham drew a detailed map of Salem village as it stood in 1692 *(figure 2),* marking the location of all households in the village, as well as some locations in the town of Salem and neighboring towns of Topsfield, Boxford, Wenham, Rowley, and Beverly. Upham used the map to support his argument that accusations of witchcraft were rooted in property disputes that had taken place years before; hence his view that the trials needed to be understood at the local village level. Like his literary predecessor, Salem-born Nathaniel Hawthorne, Upham regarded the witch trials as personal conflicts that went out of control.

In 1974, social historians Paul Boyer and Stephen Nissenbaum revisited Upham's map and used it in their study, *Salem Possessed.* They argued that the witchcraft accusations were motivated by economic and social tensions that had arisen between two factions in Salem village, one that wanted political and religious independence from the town of Salem and another that supported the town's continued governance of village affairs. The first step toward independence had been taken in 1672, when the town allowed the village to establish its own church and appoint its own minister. By 1692, however, three ministers had been appointed and dismissed. It was from within the house of the fourth, the Rev. Samuel Parris, that the first witchcraft accusations came. Boyer and Nissenbaum claimed that accusers came mainly from families who lived in the western part of the village, while the accused witches came predominantly from families living nearer to Salem town. They further identified the geographical divide with social and economic divisions in the village, arguing that families living nearer to Salem town were more closely bound to its mercantile interests and political activities. To illustrate their case, Boyer and Nissenbaum plotted accusers and accused on a map, based on Upham's, and drew a line through the center of the village to show that most accusers lived in the west and most of the accused in the east *(figure 3).* It was a powerful use of cartography to buttress theory. The authors also added a map showing the property holdings of two of the most influential families to oppose one another in the trials. Although neither of the maps prove that economic differences and conflicting loyalties to village and town caused the trials, they are strongly suggestive.

To explore the Boyer–Nissenbaum socioeconomic theory further in class, I used GIS to examine the data that they used, in addition to other data in the

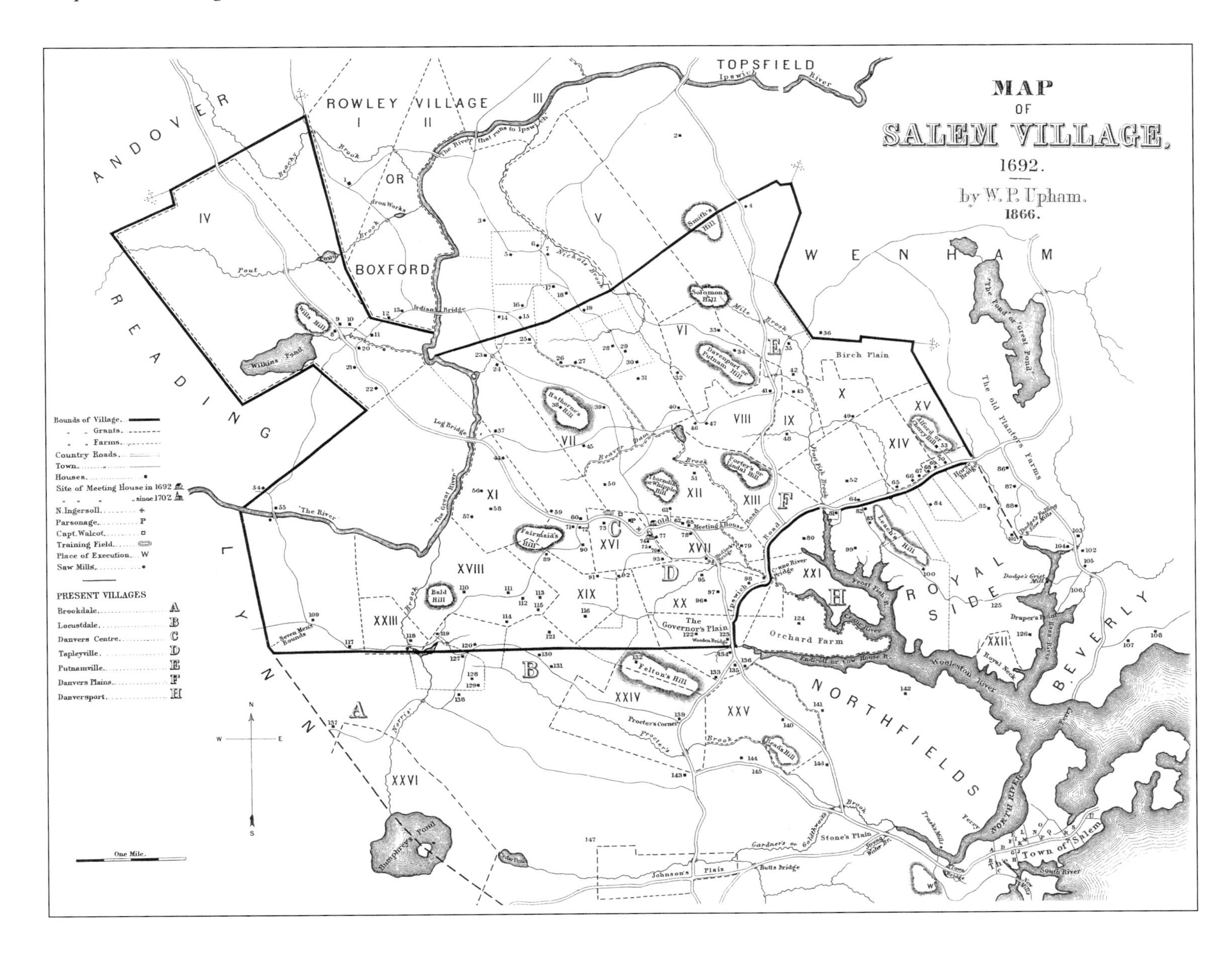

FIGURE 2. UPHAM'S MAP OF SALEM AND ENVIRONS

CHARLES UPHAM'S HAND-DRAWN MAP OF SALEM VILLAGE SHOWS THE BOUNDARIES OF LAND GRANTS AND FARMS. WHILE UPHAM MAPPED ALL HOUSEHOLDS IN THE VILLAGE, THE SCALE OF HIS MAP DID NOT ALLOW ROOM FOR MORE DETAILED MAPPING OF SALEM TOWN OR A LARGER VIEW OF THE ENTIRE REGION AFFECTED BY THE WITCHCRAFT TRIALS.

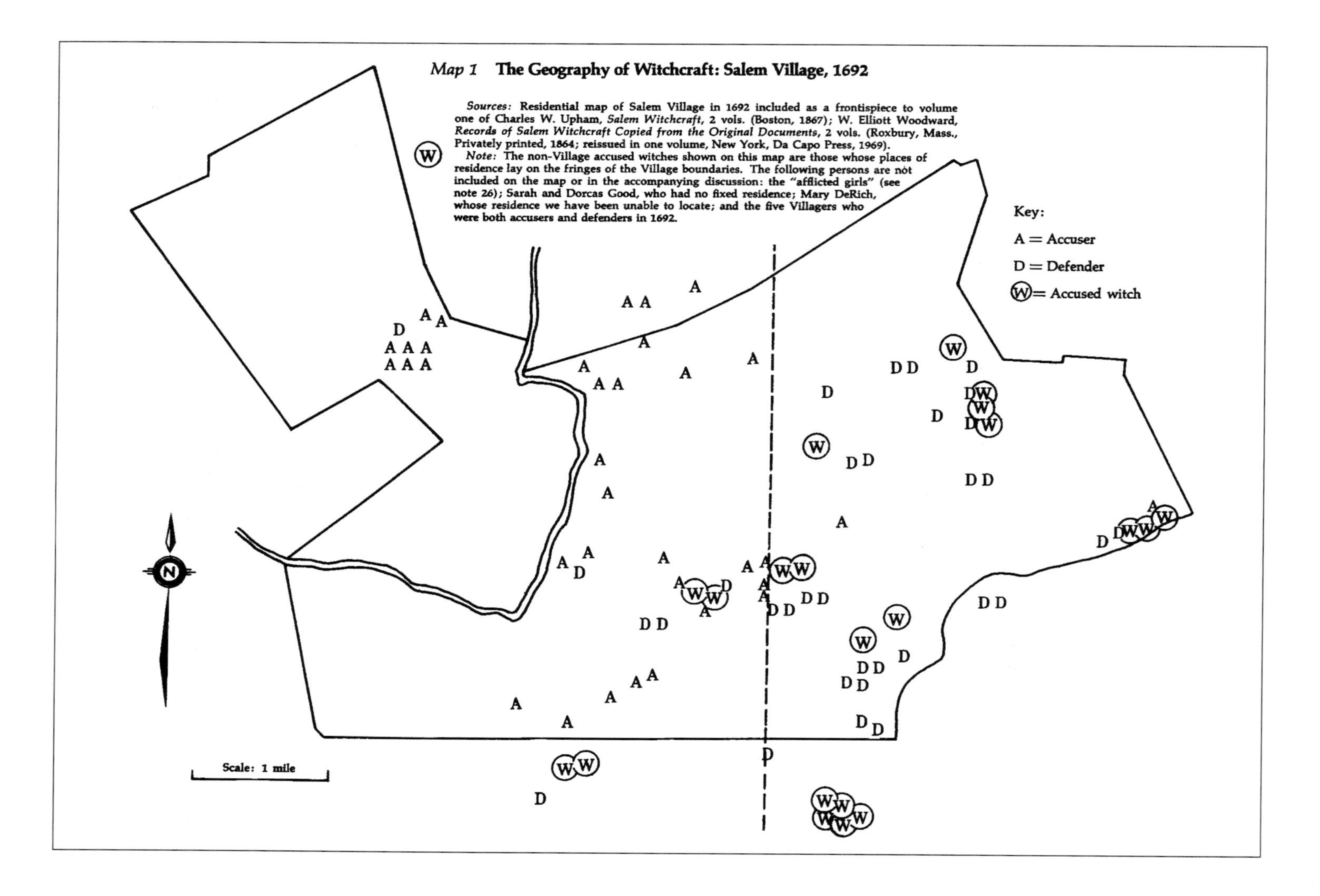

Figure 3. Boyer and Nissenbaum's map of individual accusers, accused, and defenders

This map extracts the outline of Salem village from Upham's map and adds to it letter symbols standing for the location of accusers (predominantly in the western end of the village), accused witches (mostly in the eastern village and outside its bounds), and those who defended the accused (mostly in the east). The geographical pattern seems to support the authors' argument that economic and social differences in the village lay behind the witchcraft trials. The map's generalizations and scale, however, leave out important details that support other interpretations. Reprinted by permission of the publisher of Salem Possessed, by Paul Boyer and Stephen Nissenbaum, p. 34, Cambridge, Mass.: Harvard University Press, copyright © by the President and Fellows of Harvard College.

Salem Archive. Was a geographic divide evident in the distribution of wealth (a relatively poor west and wealthy east)? Mapping village tax rates suggests not; that in fact wealth was distributed fairly homogeneously across the community. The same was true of membership in the new village church and support for Parris (*figure 4*). None of these factors show a lopsided east/west split. These patterns suggest the importance of social factors operating in the witchcraft accusations that more recent studies have emphasized.[2] The lack of strong geographical divisions lends support to the view that the witchcraft accusations arose mainly out of personal grudges, feuds within and between families, and the social dynamics at work within the circle of girls who became the chief accusers.

I also used the Salem GIS to remap the accusers, accused witches, and their defenders, incorporating some of the ideas from more recent studies of the trials. I added to the accusers the eight "afflicted" girls whose accusations accounted for the vast majority of court cases. Boyer and Nissenbaum excluded them from their map because they did not regard the girls as "decisive shapers of the witchcraft outbreak as it evolved." I counted among the accusers eight people who defended accused witches. Boyer and Nissenbaum

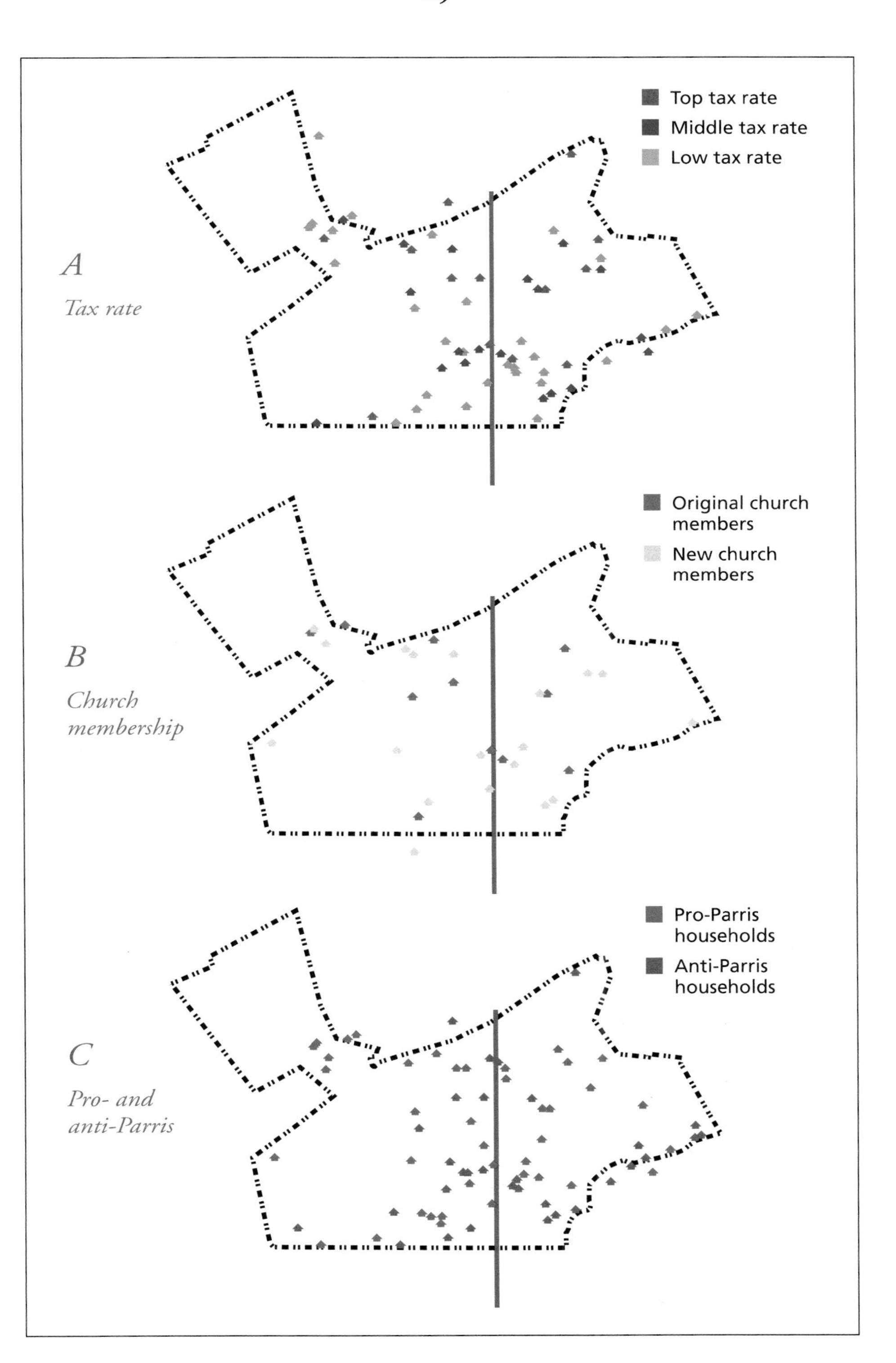

Figure 4. Mapping Salem data with GIS
Village tax rates (A), available for most village households, show a much more homogeneous spatial distribution of wealth than Boyer and Nissenbaum's argument would lead one to expect. Similarly, people from across the village joined the new church during the three years prior to the witchcraft outbreak (B), and households from all but the most distant reaches of the village opposed the Rev. Parris (C).

had excluded such people from their map. The girls fell mainly to the west of the demarcation line, while the other accusers' households were mainly to the east. Georeferencing the basemap and the line also shifted three accusers to the eastern side.[3] Despite these differences, my map shows much the same geographical division as theirs, with more accusers in the west and a disproportionate number of accused persons and defenders in the eastern part of the village.

The map of accusations can be made more revealing if we add data about social status and wealth *(figure 5).* Far more accusers than accused witches came from the households of village leaders—men who held positions in the militia, constabulary, church, and village committees during the late 1680s and early 1690s. Accusers were also more likely to come from households belonging to the top tax bracket than were witches. These findings reflect the common demographics of New England witchcraft: most people accused of witchcraft belonged to the middle or lowest social and economic brackets, rarely the top level, and few were social leaders. Economic divisions did play an important part in the witch trials, but they were not clearly expressed in the geography of settlement. These maps suggest that we should consider a wide range of divisive issues that spawned disputes within families and between neighbors—the very conflicts that are abundantly documented in the primary sources.

One of the factors that makes the Salem story different from other New England witchcraft episodes is that a large number of prominent people rose up in defense of the accused. Most of the defenders, marked by the letter "D" in figure 5, were the friends and family relations of one accused woman, Rebecca Nurse, who happened to live in the eastern side of the village. From this group came the leaders

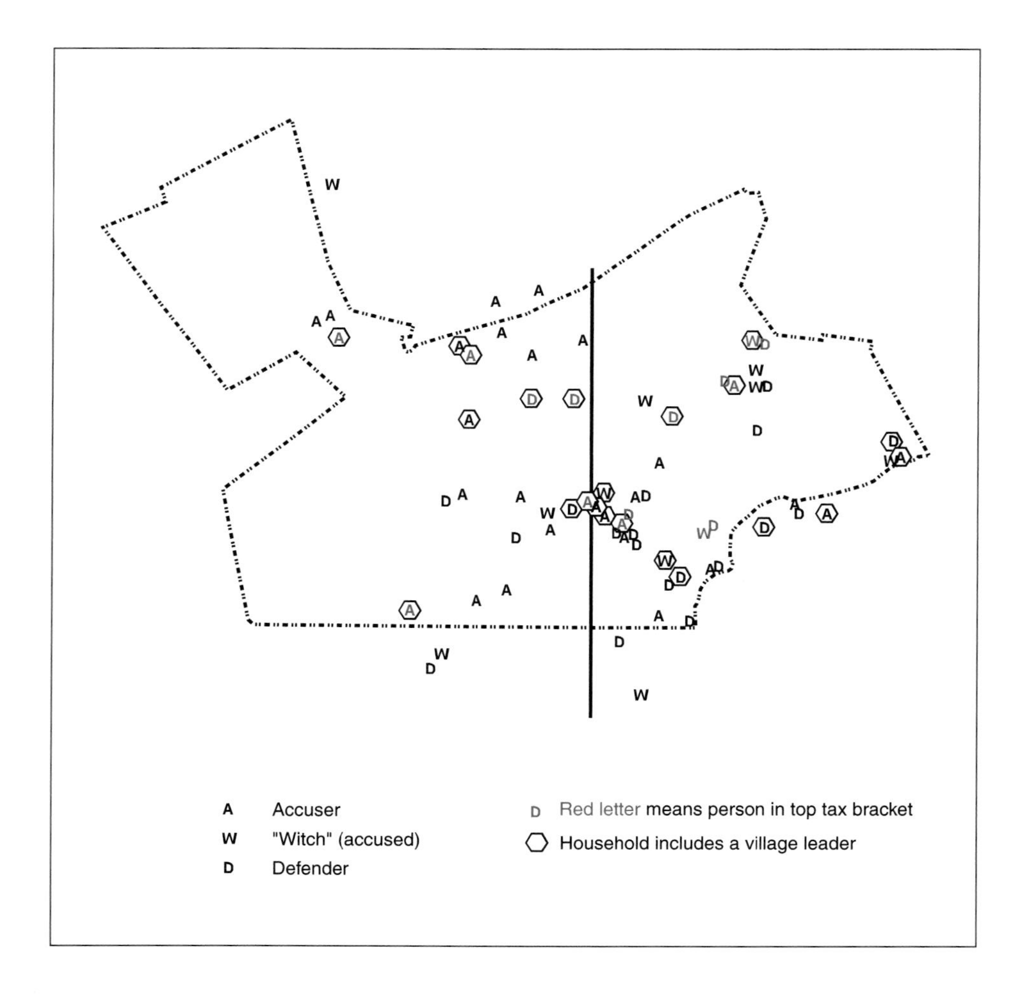

Figure 5. Households of accusers, accused, and defenders
The households of twelve village leaders included accusers (signified by the letter "A"), while only three leading households were home to accused persons (signified by "W").

of the anti-Parris faction that later ousted him. The witchcraft conflict in the village was intensified by the fact that there were village leaders on both sides of the issue.

Wanting students to explore the full range of relationships between accusers, accused, and defenders, using the documents as their guide, I asked each person in the seminar to choose two of the accused witches and to comb through the records to extract information about them, including their age, gender, family relationships, social, economic, and political position in the Bay Colony, residence, and the date and nature of their participation in the trials. With all this information organized in a relational database linked to a GIS of locations, students could investigate who was doing what to whom and study the unfolding legal process.

I also wanted to offer students a more comprehensive geographical approach. The court records include people from twenty-four different towns. Mapping the location of everyone who was involved in the trials shows that what we usually think of as the peculiar aberration of Salem in fact affected much of the eastern half of Massachusetts Bay Colony *(figure 6)*. Taking account of everyone who was involved raises questions that Perry Miller, the leading historian of Puritan New England, considered in *The New England Mind,* published in 1953. Miller believed the trials were caused by institutional failures of church and state in Massachusetts Bay. He particularly blamed the leaders of the colony, from the governor, Sir William Phips, and his close circle of advisors (including Increase and Cotton Mather) to the ministers and judges in Salem who broke judicial precedent by endorsing the use of spectral evidence and by doing nothing to dampen the accusations.

Historians generally agree that the suspension of the old Massachusetts Bay charter in 1690, and the appointment

Figure 6. Regional extent of the witchcraft trials
Salem village lay at the heart of what might better be called the witchcraft region, which by the autumn of 1692 extended from Wells, in present-day Maine, to Boston, and from the fishing port of Gloucester to Billerica. All towns shown here had one or more persons involved in the witchcraft trials by November 1692.

of Phips as governor with a new royal charter in 1692, destabilized colonial institutions. The new charter abolished the long-standing Puritan theocratic state, in which the governor, church, legislature, and courts were welded together in a single religious and civil body. The tight control formerly exercised by the colony's unified authorities came unstrung just as the witchcraft accusations began. With the courts suspended, people accused of witchcraft were charged and held in jail for up to three months pending trial. Even when Phips arrived with the charter, it was unclear how the church, state, and judiciary would now relate to each other. In this period of political uncertainty, town conflicts and personal animosities were allowed to play themselves out unchecked.

Seeing the broad geographical extent of the accusations also drew students' attention to conflicts within other towns. The class focused on Andover, the town that registered the greatest number of people accused of witchcraft and the subject of a good deal of recent research. Divisions in Andover arose between older and newer settler families and the factions that sided with them. These divisions attracted and reinforced accusations from two of the girls of Salem village, who were invited to Andover to identify witches in the community. The Salem accusers, with the help of one of the ministers, pointed to people who were already suspect in public opinion or who were in conflict with others in the village. The girls' accusations then gave license for local people openly to join in. Whole families were systematically accused and brought to trial.

A series of maps generated from the Salem GIS shows the chronology of accusations and their spread across the colony from the first court hearing on February 29, 1692, to the last day of trial on November 30, 1692. By late April, people were being accused of witchcraft in a ring of towns surrounding Salem *(figure 7)*. A rash of accusations in Andover in midsummer coincided with the Salem girls' visit. The four maps shown here depict significant phases of the process, showing leaps in the number of accusations and their widening distribution across the region. The online animated maps from which these images were taken show the pattern changing day by day. The sequence of maps shows an explosion of accusations across the landscape at the local level—a dramatic representation of the temporary breakdown of the once tightly controlled Puritan social order. It was this widening collapse of church–state order that led

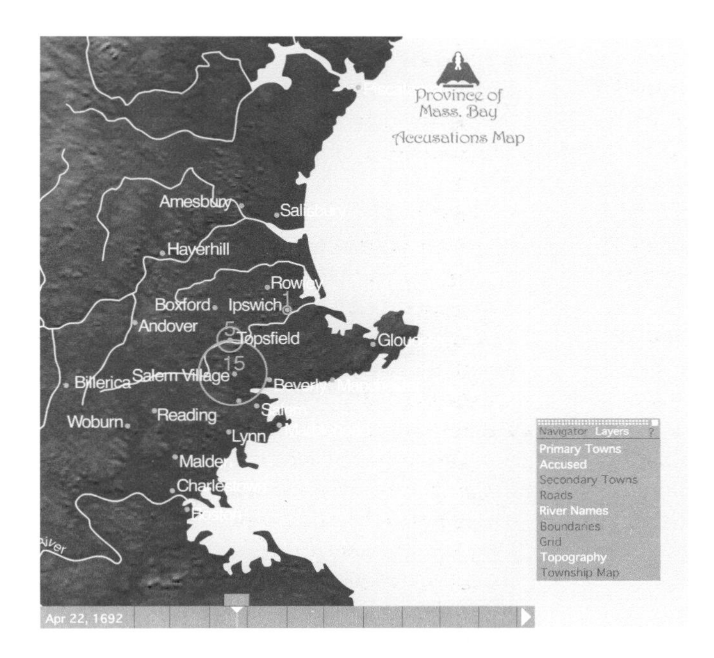

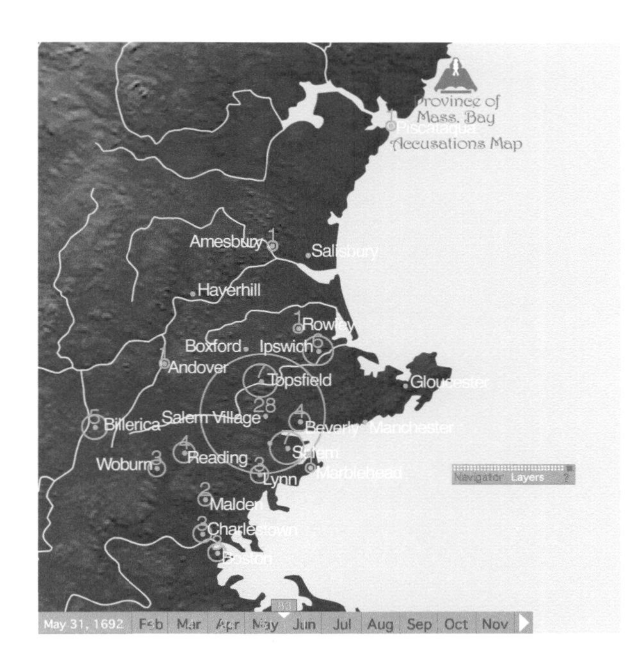

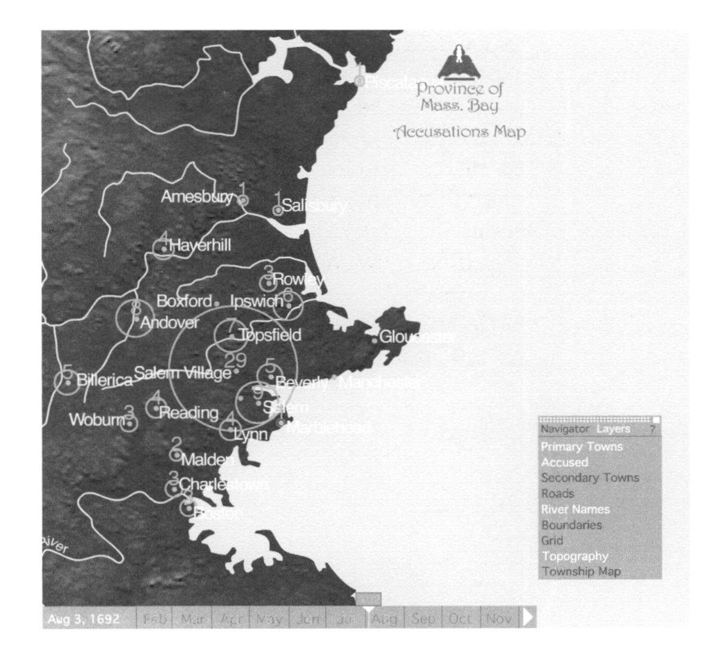

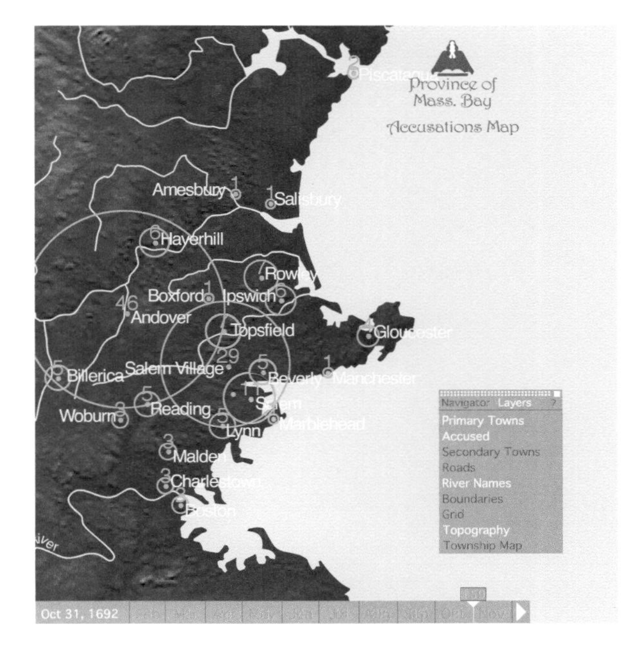

Figure 7. Spread of accusations during 1692

The first accusations of witchcraft were in Salem village and neighboring Ipswich. By the end of May, they extended as far north as Amesbury and as far west as Billerica and Woburn. After June, no one was accused in Salem, but more and more people were swept into the trials from outlying communities, particularly Andover, where forty-six people stood accused by the end of September.

to the popular New England characterization of the Salem witchcraft outbreak as a "hysteria" and a mass "delusion." The animation of this process, taken from information recorded in the court records, can be paused to note developments day by day at any point along the time line.

Seeing the explosion of accusations in Andover prompted vigorous class discussion. Did the geographical expansion of accusations signify the contagious spread of notions of witchcraft—simply the hysteria of popular belief—or were other factors at work? Because the Andover accusations involved extended families, they seemed especially calculated, not hysterical as much as personally directed. The records also reveal an interesting characteristic found in other New England episodes. Powers of witchcraft were ascribed to female bloodlines, hence the great susceptibility of women and their female siblings and offspring. Several students devoted their work to Andover cases to explore this important theme.

Because all the information in the GIS links individuals to their geographical location, students can follow the social, spatial, genealogical, and legal relationships involved in particular cases. The story of Rebecca Nurse in Salem village exemplifies many of the elements of neighborly conflict identified by Upham and Boyer and Nissenbaum. The accusations against Rebecca (Towne) Nurse, who was generally regarded as a model of Christian piety, were a turning point in the trials *(figure 8)*. If she could be accused, anyone was vulnerable. Several years prior to the witchcraft accusations, Rebecca's family, the Townes, won a series of judgments over land disputes with the Putnams, another leading family in Salem village. Rebecca's husband's family, the Nurses, prospered while the Putnams did not fare so well. The Putnams supported the new village church while Rebecca kept her membership in the mother church in Salem. When witchcraft accusations started in the village, the Putnams laid charges against Rebecca Nurse. A neighbor, Sarah Holton, added damning testimony that, three years before, Rebecca Nurse had vigorously scolded her husband, Benjamin Holton, for letting his pigs damage her vegetable garden. According to the widow Holton, Benjamin's sudden death after the incident was caused by Rebecca's witchcraft. Rebecca maintained her innocence and was put to death on the gallows.

In addition to influencing my approach to teaching history, creating the Salem Witch Trials Archive involved me in new

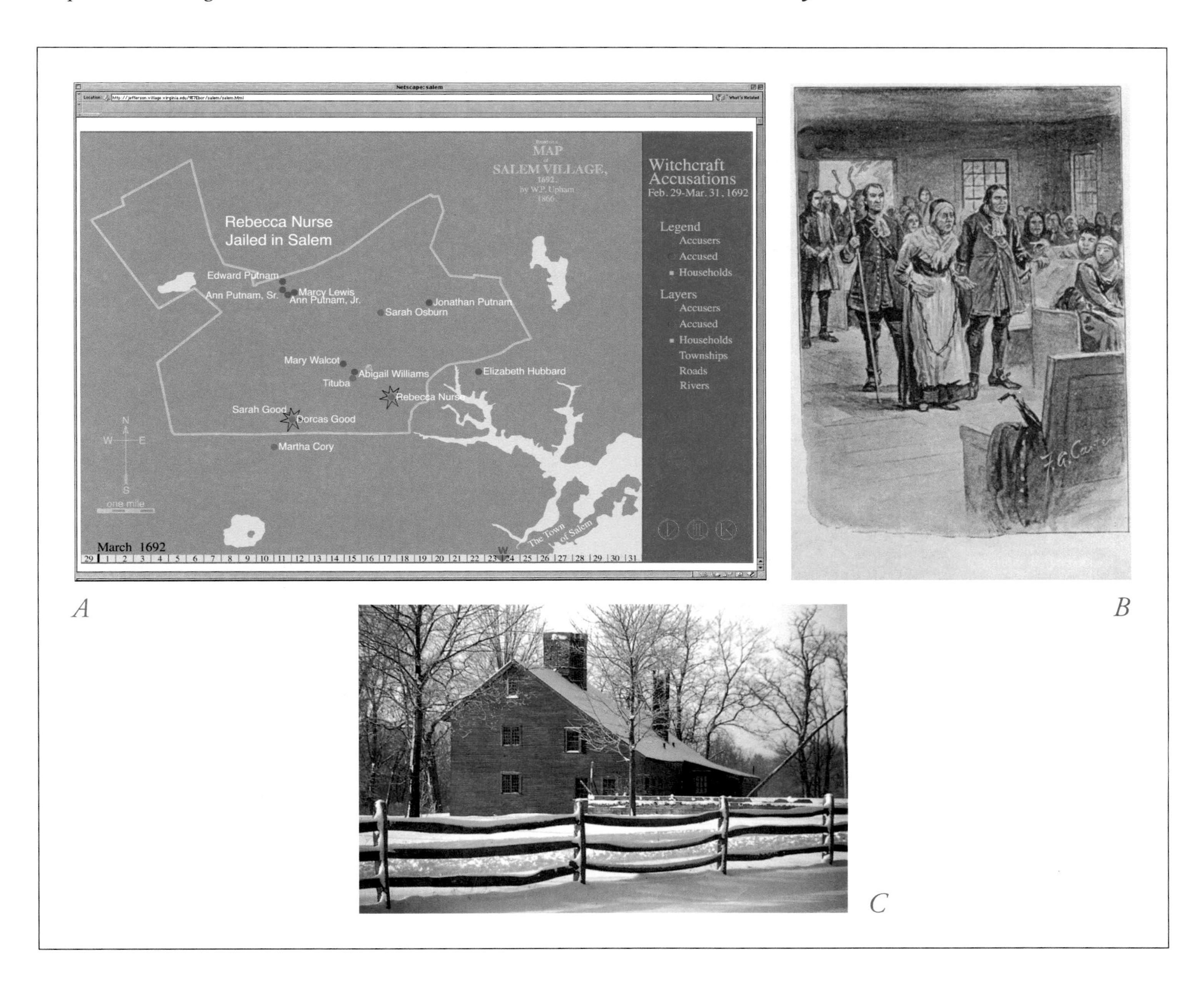

Figure 8. Visualizing the tale of Rebecca Nurse

The Salem Witch Trials Archive brings together many kinds of documents related to the trials, including maps on particular themes, artistic renderings of events, and modern images of the landscape. All images are linked by their geographical location. A search on Archive documents related to Rebecca Nurse brings up (A) a map of those involved in accusing her of witchcraft, (B) F. A. Carter's drawing of her being brought before the court, from John R. Musick's suggestively titled 1893 book, The Witch of Salem: or, Credulity Run Mad, and (C) a photograph of the Nurse house, still standing in Salem. Photograph by Richard Trask.

collaborative relationships with Salem witchcraft specialists, archivists, database designers, GIS technicians, and graphic designers. These collaborations required me to rethink the content and representation of historical source materials and how they can be used in research and teaching at various levels.

The GIS portion of the project forced me to consider for the first time how maps are made and the purposes they can serve in teaching and research. Although I could have calculated the relatively small numbers involved and plotted the data on individuals on a series of hand-drawn maps, I would not have tried to do it without GIS because of the labor and potential inaccuracies involved. Using GIS enabled me to incorporate and analyze a larger body of data, and to explore geographical patterns at a variety of temporal and spatial scales. While GIS has not changed my basic historical methods, I now routinely use spreadsheets and relational databases because they permit far more accurate and richer maps. Because all the data can be so easily shared, this approach enables—and indeed virtually demands—collaboration in both teaching and research. I am in the process of designing a new course on the Salem witch trials in which students themselves create demographic records for the maps. Finally, while GIS maps can be powerful devices for asking new questions of historical subjects, they can also become powerful visual components in developing new interpretive arguments.

Further reading

Boyer, Paul S., and Stephen Nissenbaum. *Salem Possessed: The Social Origins of Witchcraft.* Cambridge, Mass.: Harvard University Press, 1974.

Karlson, Carol. *The Devil in the Shape of a Woman.* New York: W. W. Norton, 1998; orig. pub. 1987.

Miller, Perry. *The New England Mind: From Colony to Province.* Cambridge, Mass.: Harvard University Press, 1953.

Norton, Mary Beth. *In the Devil's Snare: The Salem Witchcraft Crisis of 1692.* New York: Knopf, forthcoming.

Reis, Elizabeth. *Damned Women: Sinners and Witches in Puritan New England.* Ithaca, N.Y.: Cornell University Press, 1997.

Rosenthal, Bernard. *Salem Story: Reading the Witch Trials of 1692.* Cambridge: Cambridge University Press, 1997.

Upham, Charles W. *Salem Witchcraft: With an Account of Salem Village, and a History of Opinions on Witchcraft and Kindred Subjects.* Boston: Wiggin and Lunt, 1867.

Online resource
Salem Witch Trials Archive: www.jefferson.village.virginia.edu/salem

Notes

1. The Archive and the research supporting it remain in progress.
2. Carol Karlson, *The Devil in the Shape of a Woman* (New York: W. W. Norton, 1998; first published in 1987); Bernard Rosenthal, *Salem Story* (Cambridge: Cambridge University Press, 1993); and Mary Beth Norton, *In the Devil's Snare: The Salem Witchcraft Crisis of 1692* (New York: Alfred A. Knopf, forthcoming).
3. I was unable to identify several households at the center of Boyer and Nissenbaum's map, marked by the letters "A," "W," and "D," owing to inaccuracies of the hand-drawn map and to the fact that Boyer and Nissenbaum do not identify most of the people (and households) marked by the letters they placed on the map.

CHAPTER THREE

SIMILARITY AND DIFFERENCE IN THE ANTEBELLUM NORTH AND SOUTH

Aaron C. Sheehan-Dean

During the 1850s, Americans living in the North and South came to see themselves as different from each other. Following that cue, many historians have identified how the differences between Northern and Southern society grew more pronounced and problematic during the antebellum period. Others have seen more convergence than divergence when they analyze the values and practices of Americans in the decade before the war.[1] Researchers at the University of Virginia wanted to examine the differences attributed to regions at the local scale by comparing one Northern and one Southern county before, during, and after the Civil War. To do so, they created a digital archive called Valley of the Shadow: Two Communities in the American Civil War. It gathers a wealth of primary source material about two rural counties located at either end of the Great Valley, which straddled the North–South divide *(figure 1)*. Soldiers from Franklin County, Pennsylvania, fought for the Union, while those from Augusta County, Virginia, sided with the Confederacy. Yet the residents of these two counties had many things in common—similar cultural backgrounds and settlement histories, shared ecology and terrain, and complementary economic systems. Besides wanting to make materials about the two places available for others to study, the UVA researchers wanted to explore the similarities and differences between Franklin and Augusta counties, and the consequences of their shared or exclusive concerns for the coming of the Civil War.

Valley of the Shadow was initially composed of personal manuscripts, census rolls, and tax lists, sources historians have long used. This material was digitized to make computer analysis and online searches possible. The research team gradually realized that many of the questions they were asking of the conventional sources were geographical. For example, did wealthy families monopolize the most fertile lowland soils while poorer families were relegated to higher, less productive lands, as previous studies argued? Were settlement patterns in Northern and Southern communities markedly different, as one would expect from the emphasis historians have given to the deep impress of industry on the North and of slave agriculture on the South? How different were the two counties' transportation systems? Did neighborhoods vote in coherent blocks? Answering these questions meant adding geographical information to the archive.

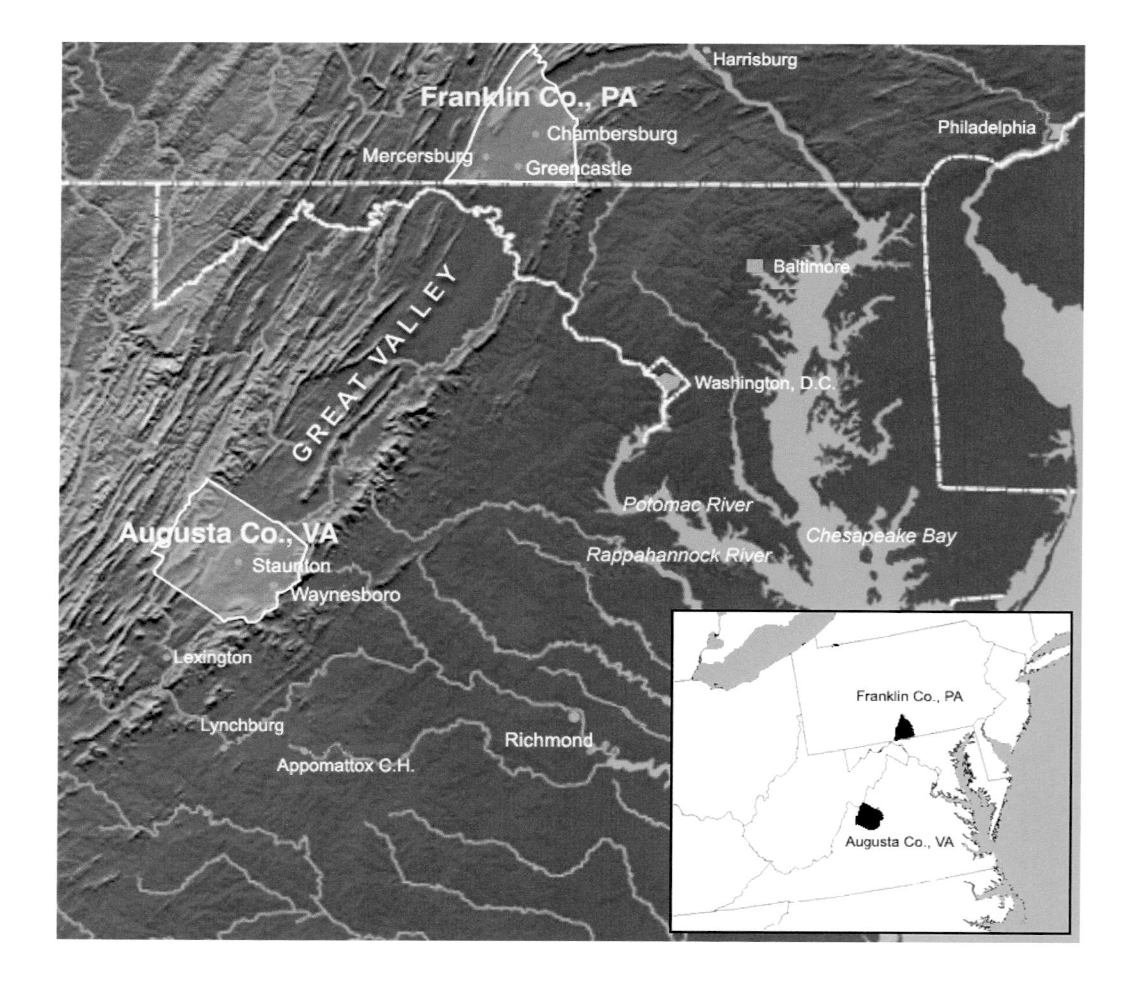

Figure 1. Franklin and Augusta counties

Both counties are located in the rolling and sometimes rugged terrain of the Great Valley and both were settled in the late eighteenth and early nineteenth centuries by immigrants from the eastern United States, Great Britain, and Germany. Franklin and Augusta were typical of antebellum border communities, with ties of kinship, culture, and trade connecting relatives, friends, and business associates across regions.

We therefore created a geographic information system (GIS) that related archival information about individuals and localities to their geographic location. The diaries and letters in the archive's manuscript collection helped suggest this approach, for they revealed that, to a great extent, the distinct perspectives of people in Augusta and Franklin were based on their engagement with hometown issues and society. Although many residents of both communities participated in national organizations and international trade networks, their outlooks reflected the intense localism typical of the mid-nineteenth century. Understanding their views and actions demanded that we pay close attention to where people lived, and that we build up from the individual and household level to make sound generalizations about each community.

We obtained detailed antebellum maps of Franklin and Augusta that showed residences and most commercial and public buildings *(figure 2)*. We scanned the maps and then georeferenced them so that they could be registered to modern maps of the two counties. This way, we could determine the actual distance between any two places on the historic maps. We could

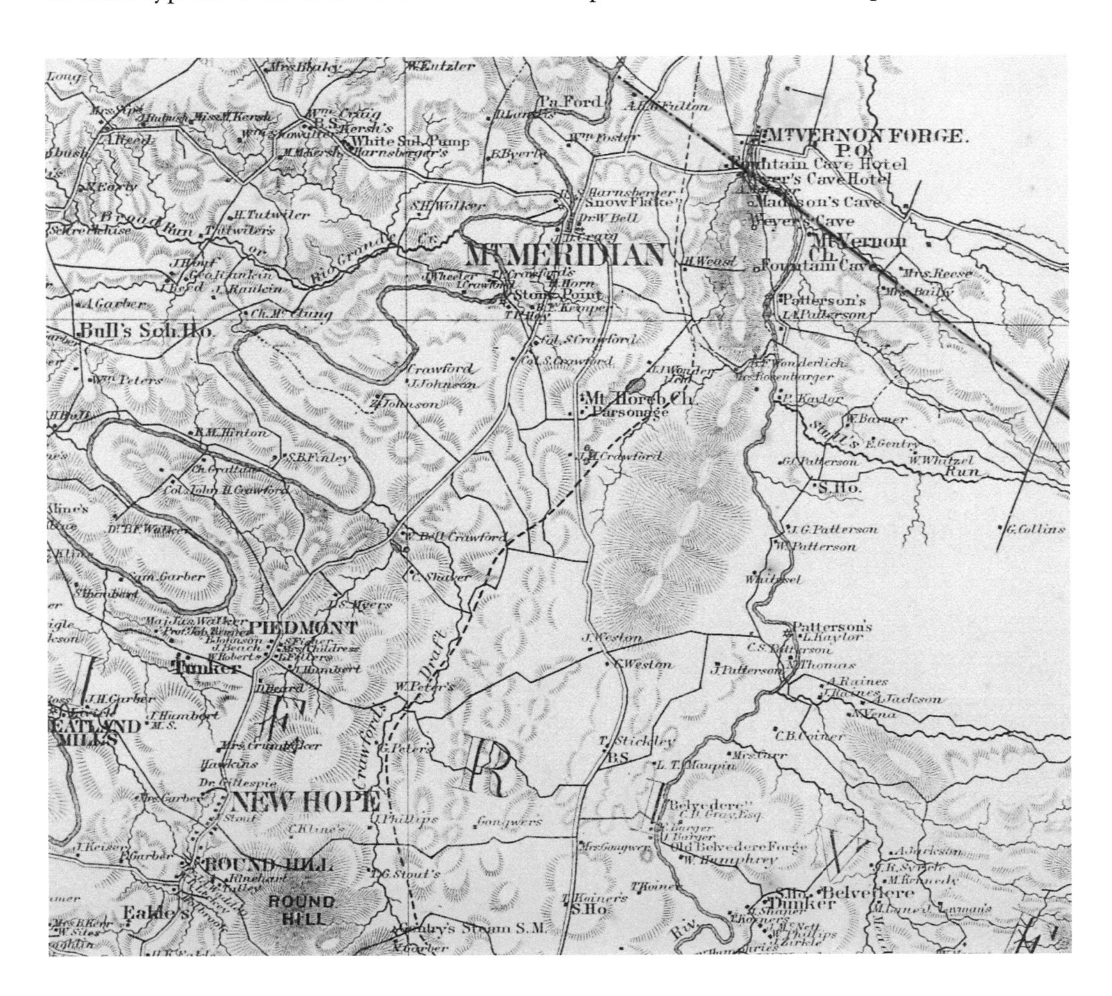

Figure 2. Eastern Augusta County
This detail is from the map that the famous Confederate cartographer Jedediah Hotchkiss drew of Augusta County in 1871. This and the 1858 Davidson map of Franklin County show mountains, streams, and valleys, as well as man-made features such as residences, public buildings, railroads, and roads. The level of detail and accuracy in both maps allowed us to identify demographic and socioeconomic information for most of the named points through the 1860 U.S. Census. There are more than four thousand named locations on the Hotchkiss map and more than seven thousand on the Franklin map. Map detail courtesy of the Special Collections Department, University of Virginia Library.

also overlay information from modern digital maps, such as elevation and soil maps produced by the U.S. Geological Survey, to examine historical places in their topographical context. Once the two maps were georeferenced, we digitized all of their features and identified as many of the places named on each map as possible. We then linked the digitized maps to data on each household, culled from the U.S. manuscript census for 1860, so that we could analyze each county's socioeconomic landscape in detail.

By 1860, the North boasted a much larger population and much denser transportation network than the South. The same was true for our sample counties. Franklin County was 20 percent smaller than Augusta, but had almost twice as many people. Franklin residents invested more heavily in the banks and development corporations that provided capital for roads, canals, and railroads, while Augusta residents invested more of their capital in land and slaves.

Did these facts illustrate the classic characteristics ascribed to North and South, Franklin County responding to the economic incentives created by dense population and Augusta County showing isolated

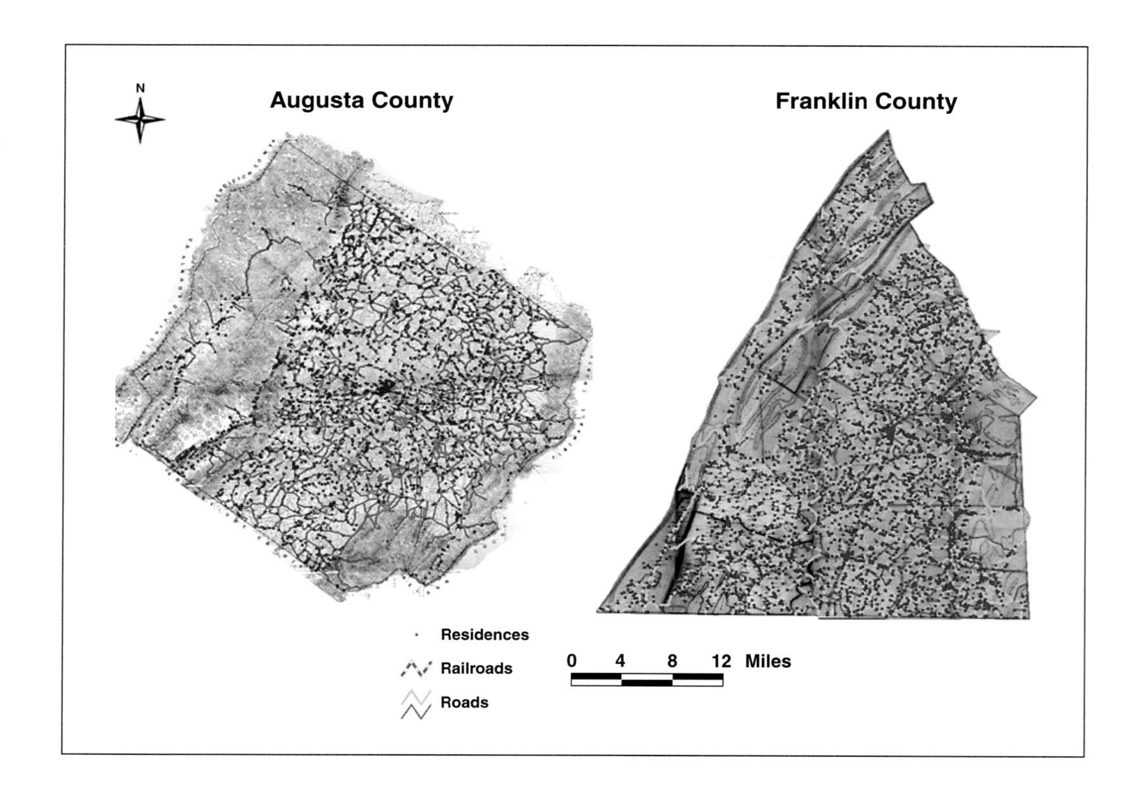

Figure 3. Railroad and road networks
Residents in both counties had easy access to the institutions and markets that sustained each community. Nearly all households in both counties, indicated by dark blue dots, were located within one mile of a major road. Franklin County, the smaller of the two in terms of land area, was more densely covered by major roads, with 1.26 miles of major road per square mile of the county compared to .53 miles per square mile for Augusta. On a per capita basis, however, Augusta had slightly more miles of road than did Franklin (.029 miles per person for Augusta and .023 miles per person for Franklin).

Southern farmers' disinterest in economic change? A glance at the antebellum maps suggests that both counties had a fairly dense network of major and minor roads *(figure 3)*. Each had a railroad running through the county seat. GIS allowed us to investigate the practical consequences of road development for residents in each county. Thinking that ease of access to good transportation routes would have been more important to antebellum farmers than the number or density of roads, we used a GIS function called *buffer* analysis to determine each household's distance from a major road. In both counties, almost every household was located within one mile of a major road. Further, although Franklin County contained more miles of major roadways per square mile of the county, Augusta held more miles of major road per person and more than twice as many miles of secondary roadway. The more dispersed settlement pattern of the Southern county did not appear to mean its farmers had inferior access to markets.

We then created one-mile buffers around each town to include all the people living in the town as well as those who lived close enough to use town services regularly *(figure 4)*. Franklin had

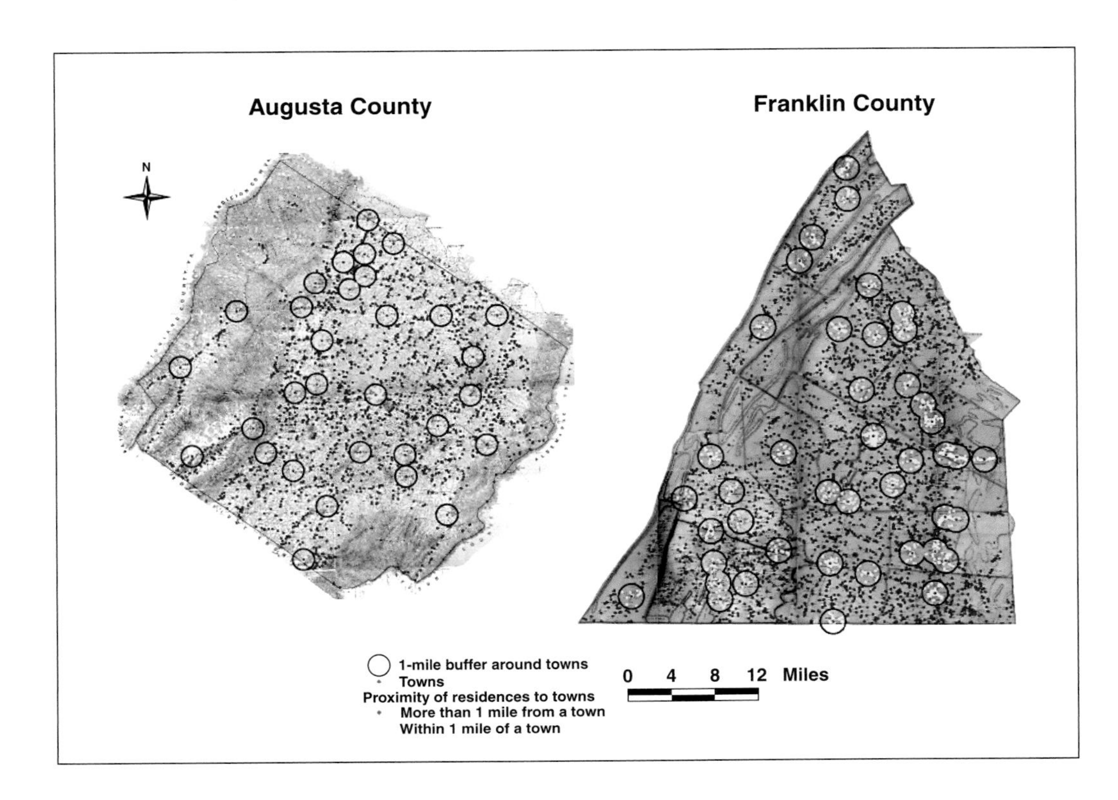

Figure 4. Residences and towns These maps show the residences in each county with one-mile buffers drawn around each town. Fifty-five percent of Franklin residents lived in towns while the comparable number for Augusta County was 30 percent. Franklin had several "second-tier" towns located in the broad, level plain south of Chambersburg (the most intensely yellow-filled circle), all of which rivaled Augusta's county seat in terms of population and level of commercial development. The residential density in the towns of each county reveals one of the important differences in Northern and Southern urbanism on the eve of the Civil War.

larger and more densely populated towns. Nearly 55 percent of its residents lived in a town compared to 30 percent of Augustans. But both counties supported two newspapers and a variety of charitable and social clubs. People in both places lived at similar distances from schools, churches, and other communal institutions. For example, the average distance from a residence to the nearest church was about 5.5 miles in Franklin and Augusta, necessitating a short buggy ride, or a long walk, to services on Sunday mornings.

In addition to providing the geographical framework we needed to create the Valley of the Shadow GIS, the antebellum basemaps provided information that had not been available, or was not so conveniently summarized, in other sources. For example, the historic maps identified and located many businesses and public buildings that gave important clues to the nature of commerce and urban life. Franklin County had one city, Chambersburg, and several large towns. On average they had five times more stores per person than did Augusta's towns, and they offered customers a much greater variety of goods *(figure 5)*. Most Franklin towns had at least one confectioner, while Augusta's only

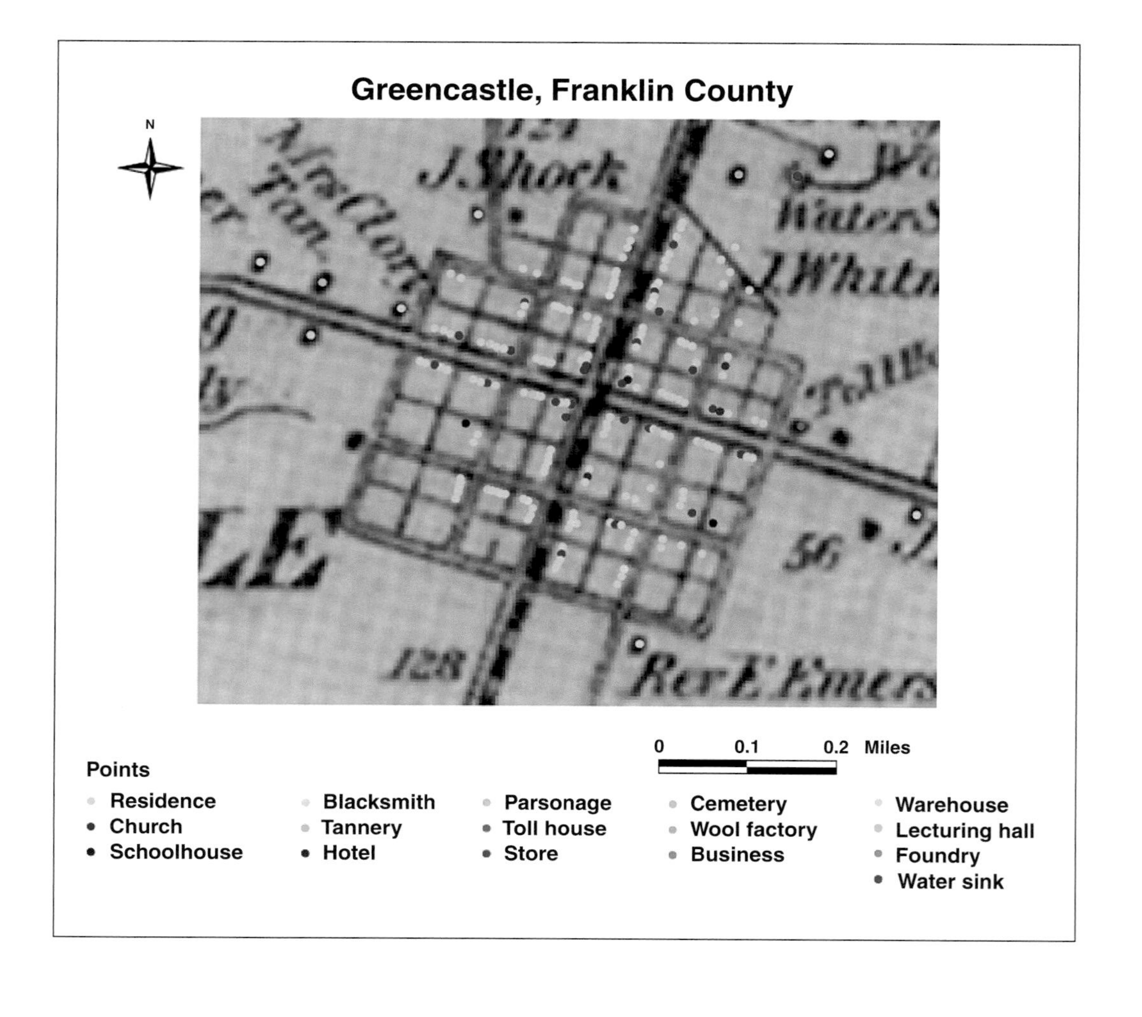

Figure 5. Greencastle, Franklin County
The Northern county held several second-tier towns, all of which were plotted on grids. In addition to having substantially higher rates of residential settlement, towns like Greencastle had a more diverse assortment of urban commercial establishments. The cooper and wagon shops were typical of mid-nineteenth-century America, but Greencastle also housed a shoe shop, a tobacconist, and a jewelry store.

sweets shop was in Staunton, the county seat. Augusta's smaller towns, many established at crossroads or natural gathering places like mountain gaps or hot springs, did not support a flourishing retail trade. But although Augusta towns had far fewer hat shops and silversmiths, they had more saw mills and grain mills. Many Augusta farmers traveled shorter distances than Franklin farmers did to convert their corn and other crops into marketable commodities. Augusta slaveholders bought most locally produced corn to feed their slaves, the excess going to one of the county's forty distilleries.[2] In this instance, the GIS helped us see how Augusta's more dispersed settlement pattern could support a vibrant economy *(figure 6)*. The geography of its infrastructure and commercial establishments supported its agricultural economy logically and efficiently.

Southern historians have analyzed the influence that slaveholders exercised over their less wealthy neighbors. Some have argued that wealthy households with slaves dominated the good soil and that poor families, especially nonslaveholders, lived on rocky, upland terrain or piney, infertile land.[3] Land distribution in the North is generally thought to have been more

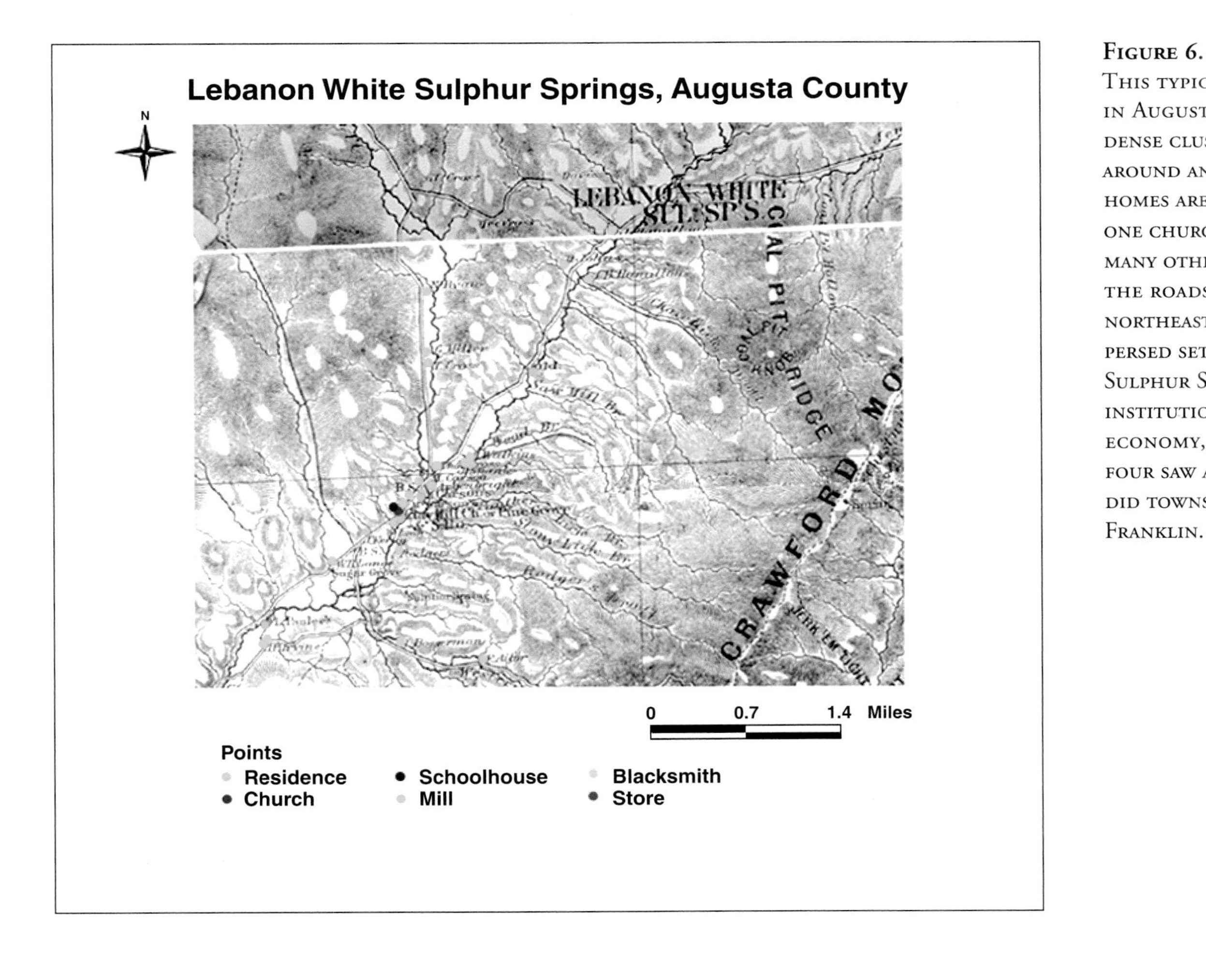

Figure 6. White Sulphur Springs
This typical second-tier town in Augusta shows a far less dense clustering of residences around an urban center. Several homes are located near the town's one church and schoolhouse but many others are located along the roads leading north and northeast. Despite this more dispersed settlement pattern, White Sulphur Springs contained more institutions supporting the rural economy, such as the town's four saw and flour mills, than did towns of comparable size in Franklin.

Figure 7. Franklin and Augusta elevation models
The central plateau of Franklin County, colored green in the image below, contained all the major towns and most of the county's homes. Although located at a higher elevation, the residents of Augusta County, like their Northern neighbors, clustered in the lower elevations. This settlement pattern, with most residences located on the more productive lower elevations regardless of wealth, was true for both counties.

egalitarian, though greater population density meant greater competition for good land. Most historians of the antebellum North argue that wealthier households responded to this competition by moving into towns, investing the profits of land sales in commerce and industry.[4] By creating digital overlays of elevation, soil type, and household wealth, we were able to test these assumptions. We also worked with each county's agricultural extension agents and other local soil experts to convert USGS soil-type categories into soil quality ratings. We identified all households by their altitude and the quality of the soil where they stood.

In both counties, elevation and soil quality appear to have had little effect on residential patterns *(figure 7)*. Although wealthy Augusta planters dominated the county's political life, they did not monopolize the best lands. Rich, middling, and poor families lived side-by-side in the most fertile areas. The few Augusta households with more than thirty slaves farmed more efficiently than anyone else in the county, as measured by bushels per acre. But nonslaveholding families farmed as efficiently as the other slaveholders, at least in part because they had access to equally productive land.

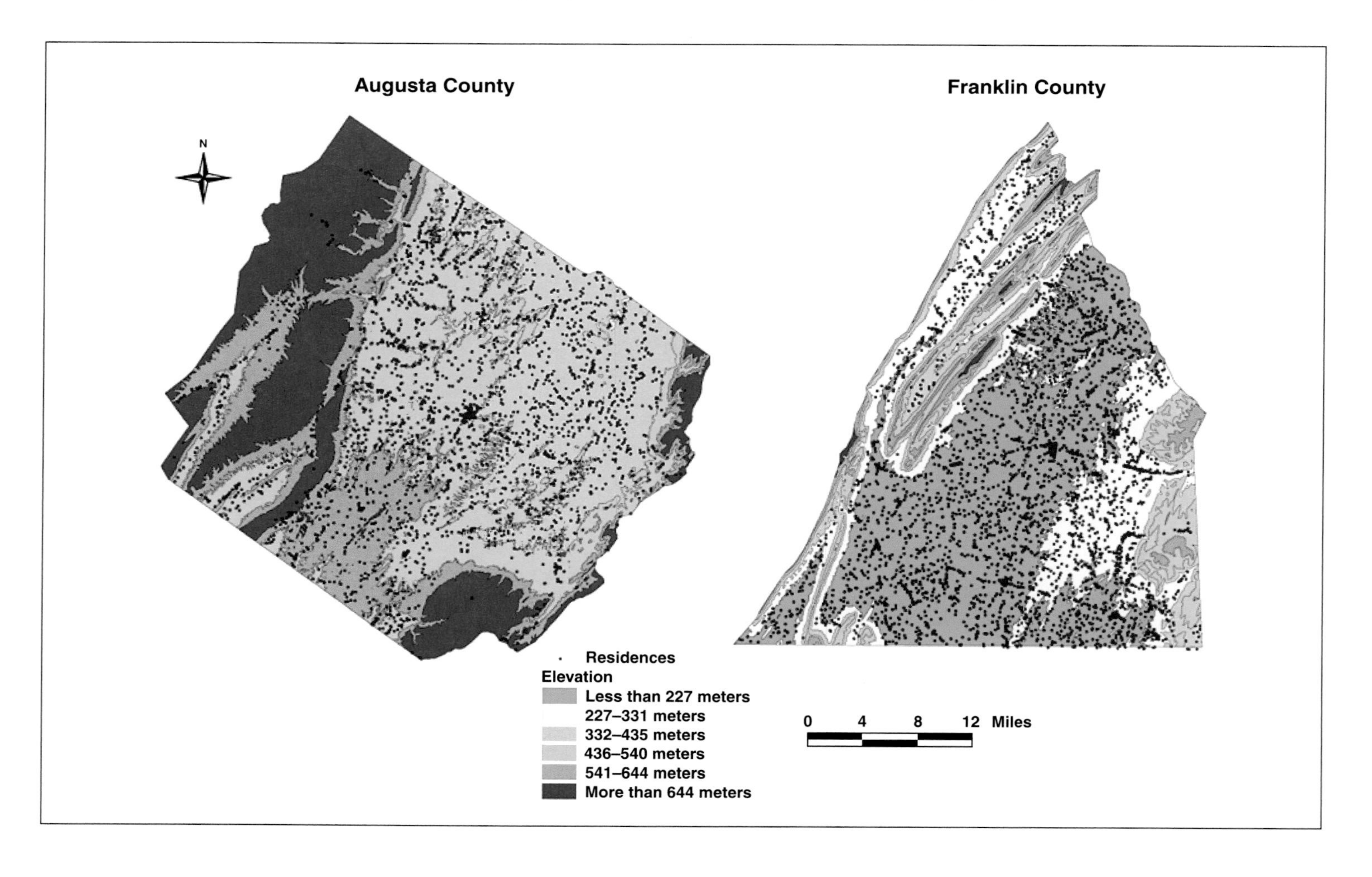

As in Augusta, Franklin families avoided higher elevations, settling mainly in the broad fertile plain south of Chambersburg. Soil quality may have made more of a difference in Franklin; farm families on the best soil invested more heavily in wheat production and generated higher profits than their neighbors on poorer soil *(figure 8).* Despite the rapid growth of Franklin's towns, the distribution of wealth shows that agriculture was still very lucrative there in 1860. The wealthiest Franklin households lived outside of towns, where they produced wheat for export to markets in Baltimore and Philadelphia.

The election of 1860 marked the first time that a Republican candidate won the presidency. We wanted to know who supported Lincoln, who sided with the moderates John Bell and Stephen Douglas in their search for sectional compromise, and who rejected conciliation in favor of John Breckinridge's radical stance. Household-level voting returns are scarce for the antebellum period. Since we knew the political affiliation of only the small number of political activists in each county, we searched for another way to analyze voting behavior below the county level.

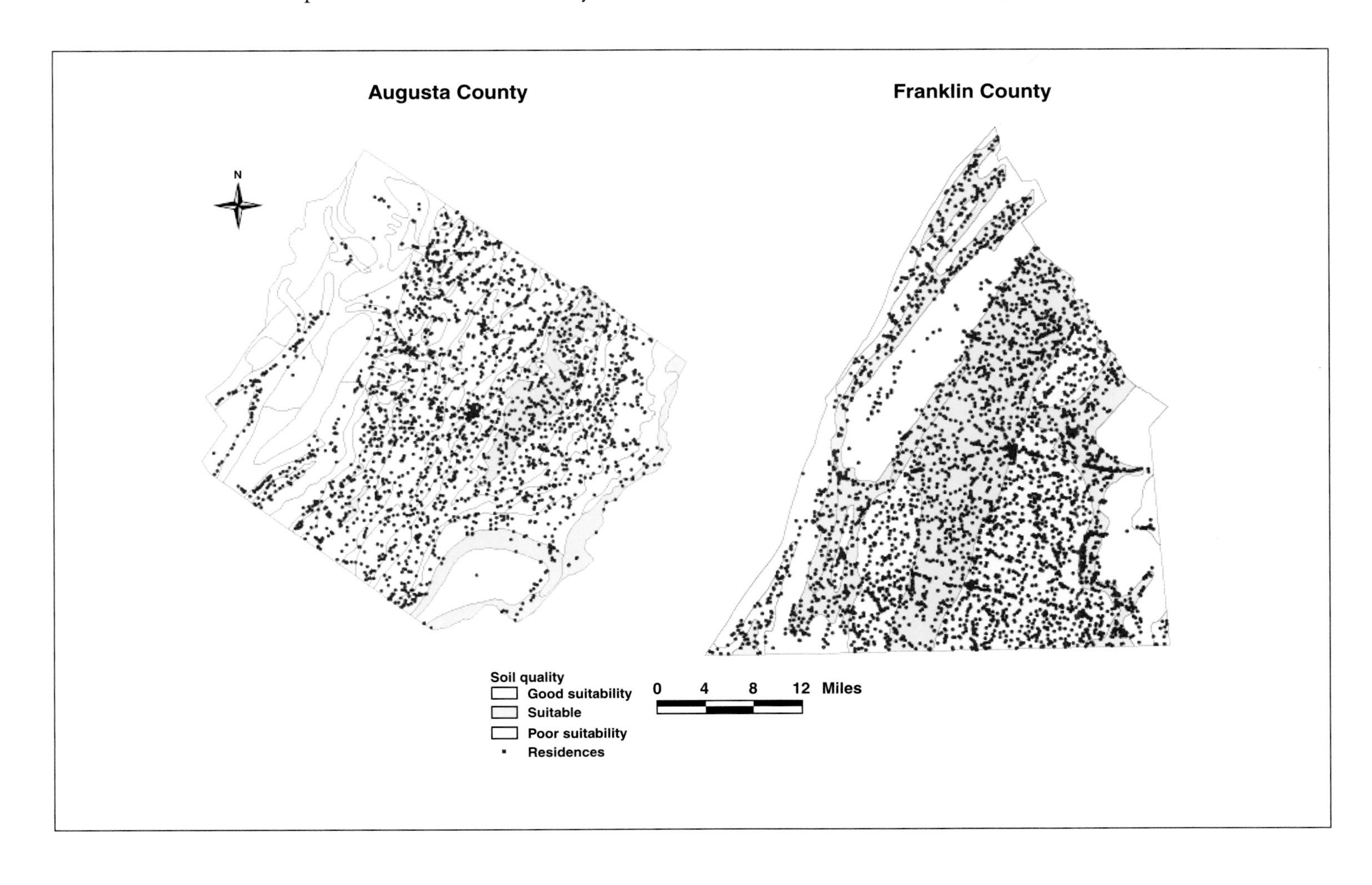

Figure 8. Franklin and Augusta soil quality
In Franklin, 95.7 percent of households lived on soils suitable for agriculture. The equivalent figure for Augusta was 91.5 percent. The GIS enabled us to compare the efficiency of farm output while holding constant variables like soil quality and elevation.

From newspaper reports of the 1860 election, we obtained lists of polling stations in each county (twenty in Augusta, twenty-three in Franklin) and created new GIS layers with the precinct polling stations as discrete points. We then created *Thiessen polygons* around each station, a procedure that bounds regions according to their proximity to selected points. The lines defining the polygons around polling stations approximated precinct boundaries, whose actual boundaries were never mapped. Once these were established, we drew on statistical analysis of the household census returns to create composite sketches of the socioeconomic characteristics of each precinct and compared those characteristics to voting returns.[5]

Our results cast doubt on the inevitability of regional conflict and on easy assumptions about who supported each candidate *(figure 9)*. The precinct-level analysis showed that although Bell, the Constitutional Union candidate, drew support from wealthy Augusta farmers and townspeople, his strongest support was in places with less than half the county's average household wealth. These voters wanted to keep open the possibility of owning slaves, as it was their best

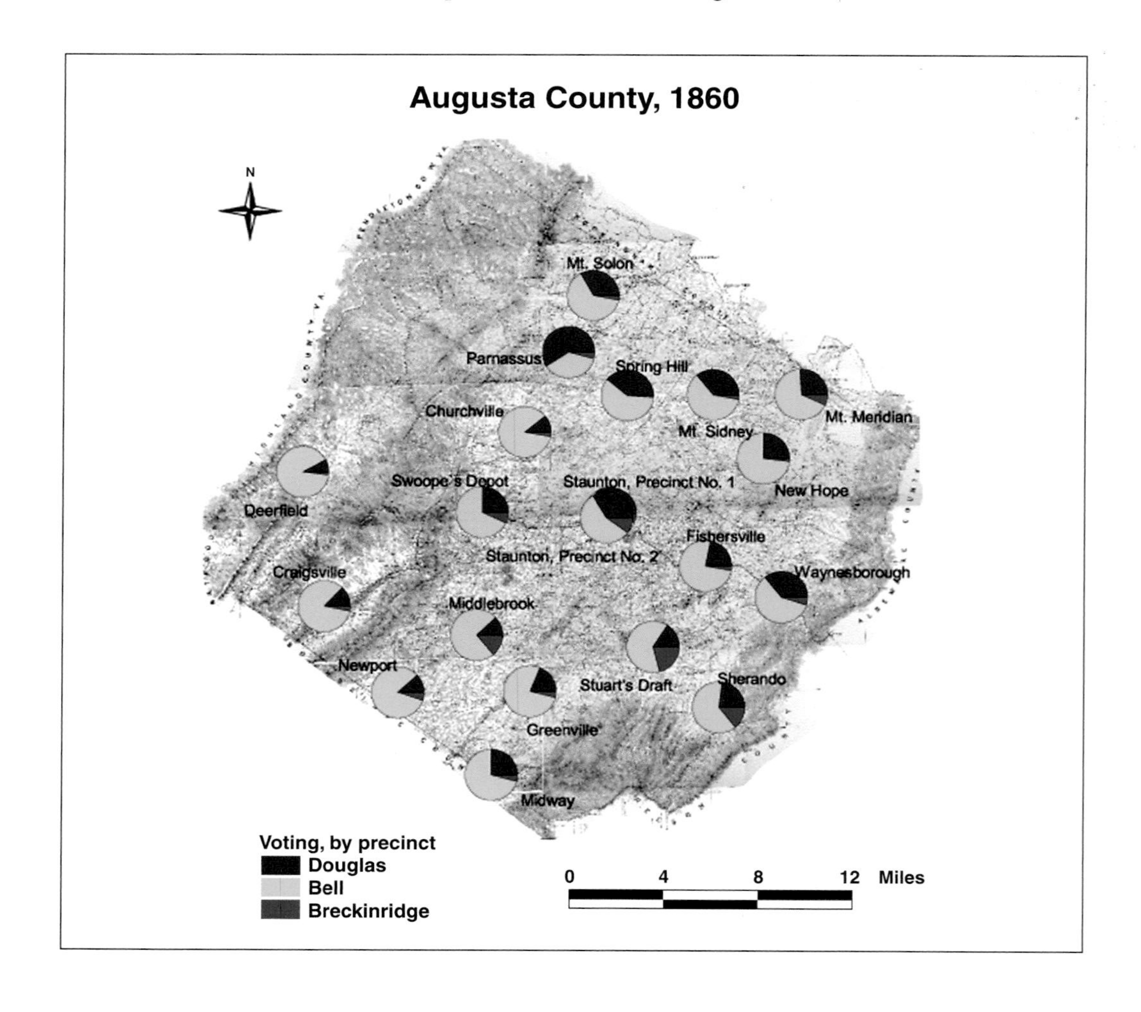

Figure 9. 1860 Presidential election returns for Augusta The moderate Whig John Bell won 66 percent, the Northern Democrat Stephen Douglas won 28 percent, and the radical Southern Democrat John Breckinridge won 5.5 percent. By creating socioeconomic and demographic profiles of voting precincts, we could examine candidates' support in finer detail than the usual county-level voting data allows.

hope for improving their economic standing, but they also felt most threatened by sectional division and the possibility of war. John C. Breckinridge, the Southern Democratic candidate, drew his strongest support in both the wealthiest and the poorest precincts. The economic security of those in wealthy precincts with high rates of slaveholding typified the strongholds of pro-slavery, anti-Lincoln voting in the South. These voters were committed to the expansion of slavery as the basis for the Southern economic system. Households in precincts giving Douglas a significant share of votes held wealth at almost exactly the county average. Bearing out the arguments of previous historians, the precincts that generated most support for Douglas had fewer slaves than did the high Breckinridge or Bell precincts. Despite Douglas's earnest efforts, Virginia voters seemed to perceive him as hostile to slavery.

As in Augusta, localized patterns of political allegiance emerged in Franklin County, perhaps reflecting the dominance of politically active or economically or socially powerful men over their neighbors *(figure 10)*. Breckinridge ran most strongly in northern precincts that were

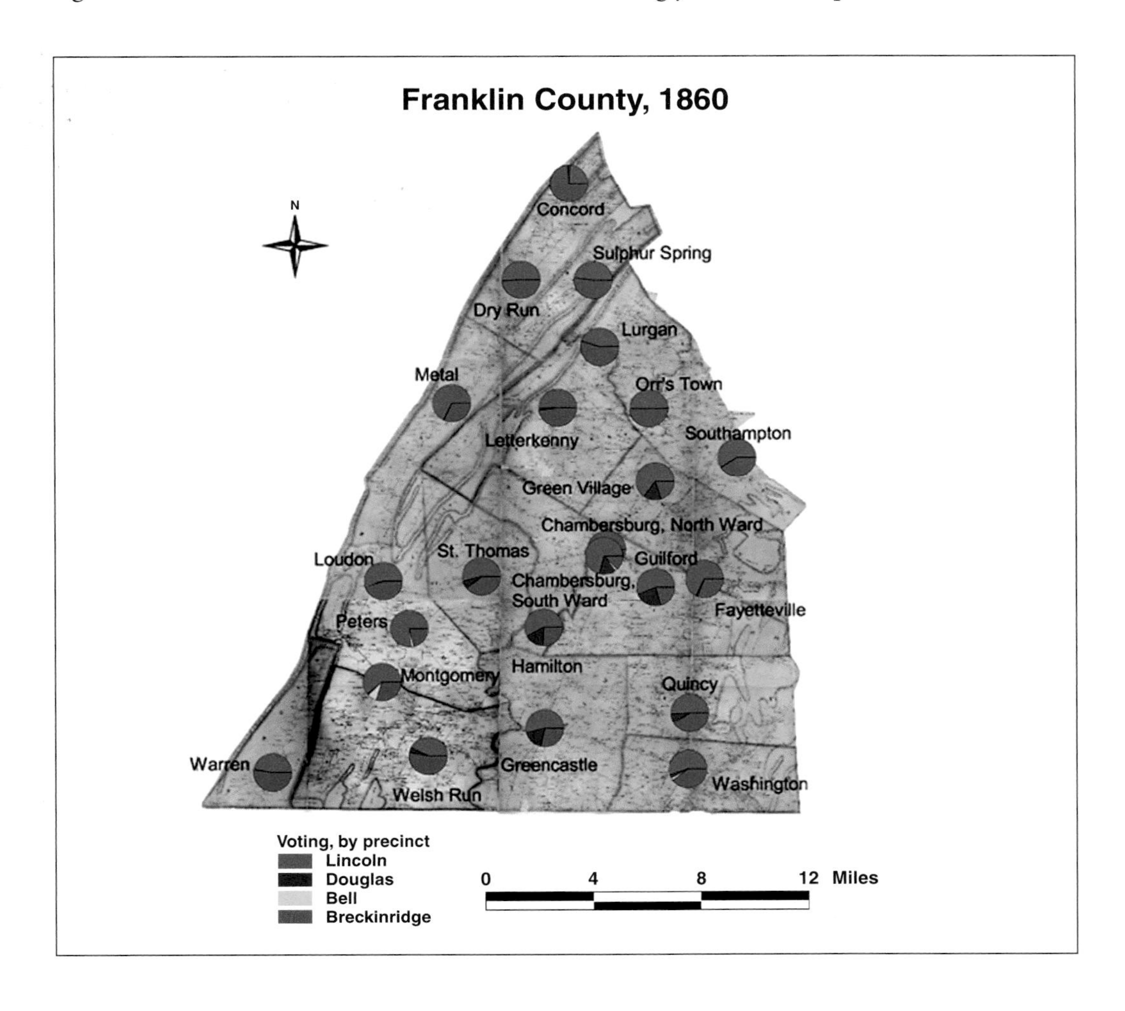

Figure 10. 1860 Presidential election returns for Franklin
As in Augusta, the presidential candidates tended to garner support from contiguous precincts, revealing the intensely local nature of most nineteenth-century politics. Republican Abraham Lincoln won 56 percent, Breckinridge 34 percent, Douglas 8.5 percent, and John Bell an inconsequential 1 percent. A last-minute reversal by Franklin County's Democratic paper to support Breckinridge helped shift Douglas votes to the pro-Southern candidate.

less wealthy than the county average and whose agricultural profile was more similar to Augusta's middling farmers—they grew more corn and less wheat than was typical in Franklin. It is unlikely that Breckinridge supporters in Franklin were deliberately associating themselves with a pro-Southern candidate because of how they farmed. Rather, with Republicans in the ascendancy, Breckinridge represented the dissenting side of Northern politics and so was the candidate of those faring less well in the fluid economic climate. Douglas won most of his votes in Chambersburg and in nearby precincts whose households held above-average wealth and farm values. He and Lincoln both did well in high wheat-growing precincts. Lincoln averaged more votes than either of his main competitors across the county and so won Franklin as a whole. The mixed returns in both counties show that these two communities, like others along the Mason-Dixon line, did not eagerly or unanimously embrace sectional politics.

The continued vitality of party politics in Franklin and Augusta illuminates one of several important similarities between these two communities that challenge arguments about broad sectional differences. In terms of settlement, infrastructural development, patterns of wealth distribution, and some aspects of politics, Franklin and Augusta were close kin. But our investigation revealed important differences that distinguished the counties as well. In particular, the more diverse and intensive urbanization of Franklin's places held the potential to generate increasingly significant divergences between the two counties over time. Using the GIS to integrate geographic, social, economic, and political information about these places helped us see that the antebellum experiences of Franklin and Augusta converged and diverged simultaneously. The community-level perspective that we adopted helped us capture the complexity of life in both places and reconcile what may have appeared as contradictions at a regional level. Our experience suggests that local-scale investigations built upon a diverse base of evidence can help us understand the nature and consequences of the shared and divergent experiences of the antebellum North and South.

Map sources

Figure 1
U.S. Geological Survey, *1-degree digital elevation models,* Reston, Va.: U.S. Geological Survey, undated; DEMs from 76 degrees west and from 41 to 37 degrees north were stitched together to create the composite image. U.S. Geological Survey, *digital line graphs,* Reston, Va.: U.S. Geological Survey, undated. Scale: 1:100,000.

Figure 2
Hotchkiss, Jedediah. *Map of Augusta County, Virginia,* 1870. Philadelphia: Worley and Bracher. Scale: 1:63,360.

Figure 3
Hotchkiss, *Map of Augusta County, Virginia;* D. H. Davidson, *Map of Franklin County, Pennsylvania.* Greencastle, Penn.: Riley and Hoffman, 1858. Scale: 1:47,520.

Figure 4
Hotchkiss, *Map of Augusta County, Virginia;* Davidson, *Map of Franklin County, Pennsylvania.*

Figure 5
Davidson. *Map of Franklin County, Pennsylvania.*

Figure 6
Hotchkiss. *Map of Augusta County, Virginia.*

Figure 7
U.S. Geological Survey. *7.5-minute digital elevation models.* Reston, Va.: U.S. Geological Survey, undated. Scale: 1:24,000. For Franklin, the following DEMs were stitched together to create the composite image: Mercersburg, McConnellsburg, Fannettsburg, Shade Gap, Doylesburg, Roxbury, Shippensburg, St. Thomas, Chambersburg, Scotland, Caledonia Park, Iron Springs, Waynesboro, Greencastle, Williamson. For Augusta, the following DEMs were stitched together to create the composite image: Reddish Knob, West Augusta, Deerfield, Craigsville, Elliott Knob, Augusta Springs, Brownsburg, Greenville, Vesuvius, Big Levels, Sherando, Stuarts Draft, Waynesboro West, Waynesboro East, Crimora, Ft. Defiance, Staunton, Churchville, Stokesville, Parnassus, Briery Branch, Mt. Sidney, Grottoes. Residential data derived from Hotchkiss, *Map of Augusta County, Virginia;* Davidson, *Map of Franklin County, Pennsylvania.*

Figure 8
U.S. Department of Agriculture, Soil Conservation Service, Pennsylvania State University, College of Agriculture, Pennsylvania Department of Environmental Resources, State Conservation Commission. *General Soil Map, Franklin County, Pennsylvania,* undated. Scale: 1:253,440. Franklin soil analysis provided by Scott Metzger, Franklin County Conservation District.

U.S. Department of Agriculture, Soil Conservation Service, Forest Service, and Virginia Polytechnic Institute and State University. *General Soil Map, Augusta County, Virginia,* 1978. Scale: 1:253,440. Augusta soil analysis provided by Tom Stanley, Augusta County Extension Agent.

Residential data derived from Hotchkiss, *Map of Augusta County, Virginia;* Davidson, *Map of Franklin County, Pennsylvania.*

Continued

Map sources (continued)

Figure 9
Hotchkiss. *Map of Augusta County, Virginia.*

Polling station list and voting returns taken from *The Staunton Spectator,* November 13, 1860, page 2, column 3.

Figure 10
Davidson. *Map of Franklin County, Pennsylvania.*

Polling station list and voting returns taken from the *Valley Spirit,* November 14, 1860, page 4, column 1.

Acknowledgments

Hardware: The Valley of the Shadow GIS was constructed in the Geospatial and Statistical Data Center (Geostat) in Alderman Library at the University of Virginia. Most of the digitizing was completed on Dell™ Optiplex™ Gxi machines on a UNIX® platform. We created screenshots and performed GIS analysis such as buffering on Geostat machines running on the Microsoft® Windows NT® platform.

Software: The GIS for both counties was constructed using ArcInfo™ software. Subsequent GIS manipulation and analysis was conducted using ArcView software. All data analysis was done using the Statistical Package for the Social Sciences, with occasional use of Microsoft Excel and Microsoft Word to facilitate the transfer of data sets between formats.

Research was made possible by generous individual donors to the Virginia Center for Digital History at the University of Virginia. Edward Ayers and William Thomas conceived the idea of creating the Valley of the Shadow archive and the accompanying GIS. Scott Crocker, Steve Thompson, and Ariel Lambert did much of the digitizing, and Mike Furlough, Blair Tinker, and Patrick Yott of the Geospatial and Statistical Data Center offered crucial technical assistance.

Further reading

Bateman, Fred, and Thomas Weiss. *A Deplorable Scarcity: The Failure of Industrialization in the Slave Economy.* Chapel Hill, N.C.: University of North Carolina Press, 1981.

Bourke, Paul, and Donald Debats. *Washington County: Politics and Community in Antebellum America.* Baltimore and London: The Johns Hopkins University Press, 1995.

Crofts, Daniel. *Reluctant Confederates: Upper South Unionists in the Secessionist Crisis.* Chapel Hill, N.C.: University of North Carolina Press, 1993.

Fogel, Robert. *Without Consent or Contract: The Rise and Fall of American Slavery.* New York: Norton, 1989.

Genovese, Eugene. *The Political Economy of Slavery: Studies in the Economy and Society of the Slave South.* New York: Vintage, 1967.

Genovese, Eugene. *The Slaveholders' Dilemma: Freedom and Progress in Southern Conservative Thought, 1820–1860.* Columbia, S.C.: University of South Carolina Press, 1992.

Holt, Michael F. *The Political Crisis of the 1850s.* New York: W. W. Norton, 1992.

Oakes, James. *The Ruling Race: A History of American Slaveholders.* New York: Vintage, 1983.

Oakes, James. *Slavery and Freedom: An Interpretation of the Old South.* New York: Knopf, 1990.

Wright, Gavin. *The Political Economy of the Cotton South: Households, Markets, and Wealth in the Nineteenth Century.* New York: Norton, 1978.

Online resource
Valley of the Shadow Web site: www.jefferson.village.virginia.edu/vshadow2

Notes

1. See the classic works listed in the "Further reading" section. For two important comparative studies, see Edward Pessen, "How Different from Each Other Were the Antebellum North and South?" *American Historical Review* 85 (December 1980): 1,119–49, and John Majewski, *A House Dividing: Economic Development in Pennsylvania and Virginia Before the Civil War* (Cambridge: Cambridge University Press, 2000).

2. The evidence for this conclusion comes from an analysis of the manufacturing and household census schedules for Augusta, but as with much of the statistical data, linking it into the GIS enabled us to see connections in a much more concrete fashion than was possible through manipulation of the census numbers alone.

3. J. William Harris, *Plain Folk and Gentry in a Slave Society: White Liberty and Black Slavery in Augusta's Hinterlands* (Middletown, Conn.: Wesleyan University Press, 1985); Stephanie McCurry, *Masters of Small Worlds: Yeoman Households, Gender Relations, and the Political Culture of the Antebellum South Carolina Low Country* (New York: Oxford University Press, 1995); Paul H. Buck, "The Poor Whites of the Ante-Bellum South," *American Historical Review* 31 (October 1925): 41–54; Ulrich B. Phillips, "The Origin and Growth of the Southern Black Belts," *American Historical Review* 9 (July 1906): 798–816.

4. Mary Ryan, *Cradle of the Middle Class: The Family in Oneida County, New York, 1790–1865* (Cambridge: Cambridge University Press, 1981); Paul E. Johnson, *A Shopkeeper's Millennium: Society and Revivals in Rochester, New York, 1815–1837* (New York: Hill and Wang, 1978); Philip Shaw Paludan, *A People's Contest: The Union and the Civil War, 1861–1865,* 2d ed. (Lawrence, Kans.: University of Kansas Press, 1996); Eric Foner, *Free Soil, Free Labor, Free Men: The Ideology of the Republican Party Before the Civil War,* 2d ed. (New York: Oxford University Press, 1995).

5. As noted in the text, the voting precincts for both counties only approximate the historical ones, for which no geographic records remain. Further, we averaged household-level data to create socioeconomic and demographic portraits of each precinct. This process does reduce the precision with which we can speak about the support each candidate received. Nonetheless, since the Thiessen procedure approximates the method that voting officials applied in drawing precincts (which was to locate polling stations at population centers), we feel confident that the GIS precincts closely resemble the historical ones and that the results of our political analysis offer reliable insight into the nature of antebellum politics.

CHAPTER FOUR

TELLING CIVIL WAR BATTLEFIELD STORIES WITH GIS

David W. Lowe

SOLDIERS march down dusty roads, across fields, through woodlots. Artillery thunders. Trees are cut down, earthworks built in haste. Men fix bayonets to their muskets and march forward to the attack. Then, sometimes only hours later, the armies march on, leaving behind devastation. But time softens the landscape, concealing under leaves and dirt what happened there. What remains of that day, if it can be found, can be used to reconstruct events. The terrain itself—roads, dwellings, fence lines, and other traces of the past—can, if preserved, evoke the memory of soldiers who fought and died there. Among the many missions of the National Park Service is this one, to keep American history alive by preserving the physical remains of the past. Any survival of a battlefield can be a means for telling the story of its war.

The National Park Service's battlefield parks preserve the vestiges of combat and interpret their significance for the public. Both activities rely on maps, whether in making the case for preservation or helping visitors imagine what happened. Most national battlefields and military parks are small and lack the budget or the people to produce accurate maps or to update those they already have. For more than eight years, the Cultural Resources GIS facility, a technical assistance office within the Park Service, has located and mapped historical features for these parks, using Global Positioning System (GPS)[1] technology and geographic information systems (GIS). It has also tried to demonstrate how useful these technologies are for integrating a wide array of park activities.

"Cultural resources" are simply man-made features of the landscape that have historical or social significance. Many of the large national parks and wilderness areas have used GPS and GIS for years to generate maps of natural resources—wetlands and wildlife habitats, for example. Often spatial data for natural resource projects can be derived from satellite imagery and aerial photography available on the Internet. In such images, however, it is difficult to pick out a mossy tumbled-down stone fence, a soldier's foxhole, or a crumbling chimney that marks an old house. To map historic survivals on a battlefield, researchers first cull through old maps, battle accounts, deeds, photographs, and other documents to determine how the landscape appeared at the time of the battle. From these, they develop lists of probable survivals, which they take into the field on what amounts to a scavenger hunt.

Earthen fortifications built by the soldiers may be found on many Civil War battlefields *(figure 1).* Some trench lines snake back through the woods for miles. Before GPS, which uses a handheld computer to generate spatial coordinates from satellite signals, only expensive, line-of-sight survey techniques were available for mapping these fortifications. As a result, they were rarely mapped. With a GPS receiver, a surveyor can map a trench simply by walking along it. One can also add details to the map, such as the locations of artillery, command bunkers, or even the foxholes of individual soldiers. By knowing more precisely where the trenches of the opposing sides are located, and their size and extent, the historian can often gain insights into how a battle was fought. As important, resource managers can better plan how to preserve fragile evidence of the passing of the armies.

Figure 1. Typical Civil War earthwork or trench
This line, originally about five feet high, has eroded over the years. Soldiers made this earthwork by excavating a ditch and piling the earth into a mound, called a parapet. They could complete a trench like this in just a few hours. Some more permanent earthworks were twelve to fifteen feet high. National Park Service photo.

Cultural Resources GIS conducted one of its first projects at Fredericksburg-Spotsylvania National Military Park in Virginia in 1994–1995. This park consists of four units that commemorate the battles of Fredericksburg (1862), Chancellorsville (1863), Wilderness (1864), and Spotsylvania Court House (1864). Although fortifications played a pivotal role in all four battles, they had never been placed on a modern map. Shortly after the Civil War, however, army topographical engineers mapped the battlefields at scales of three or four inches to the mile. These military surveyors used a theodolite (an early transit) and measuring chains to capture what proved to be a fairly detailed and accurate depiction of the terrain and its fortifications.

We compared the historic maps with modern maps and selected points that appeared to correspond between past and present, such as road intersections and houses. Surveyors visited these sites and captured their spatial coordinates with GPS *(figure 2)*. We scanned the historic maps and assigned coordinates to corresponding pixels in the digital image.[2] We then georeferenced each map, interpolating coordinates for every feature drawn by the military cartographers and making it possible later to add digital layers of modern roads, streams, and contour intervals. Armed with these maps, the surveyors returned to the field. Despite the notorious Virginia humidity, heavy undergrowth, and hungry ticks, three two-person survey teams took just twenty-five days to complete a GPS inventory of more than thirty-five miles of trenches *(figure 3)*. While recording the size and condition of earthworks, we produced the first detailed maps of the park's hiking trails, monuments, and interpretive signs. We were surprised at our ability to navigate the terrain using the historic maps alone. They often led us to survivals in deep woods where we would not have thought to look.

We intended to make this information available to park personnel but the park lacked funding to employ a GIS specialist. For several years, the data languished on a hard drive. It was easier to plod around mapping things than to place data where it would be most useful—on the desktops

Figure 2. GPS Surveyor James Stein mapping Civil War trenches
The parapet and water-filled ditch are just behind him. While collecting signals with a backpack antenna and receiver, he enters descriptive information about the resources in the handheld computer. The information is used to document condition and threats, to plan for protection, and to schedule maintenance. National Park Service photo.

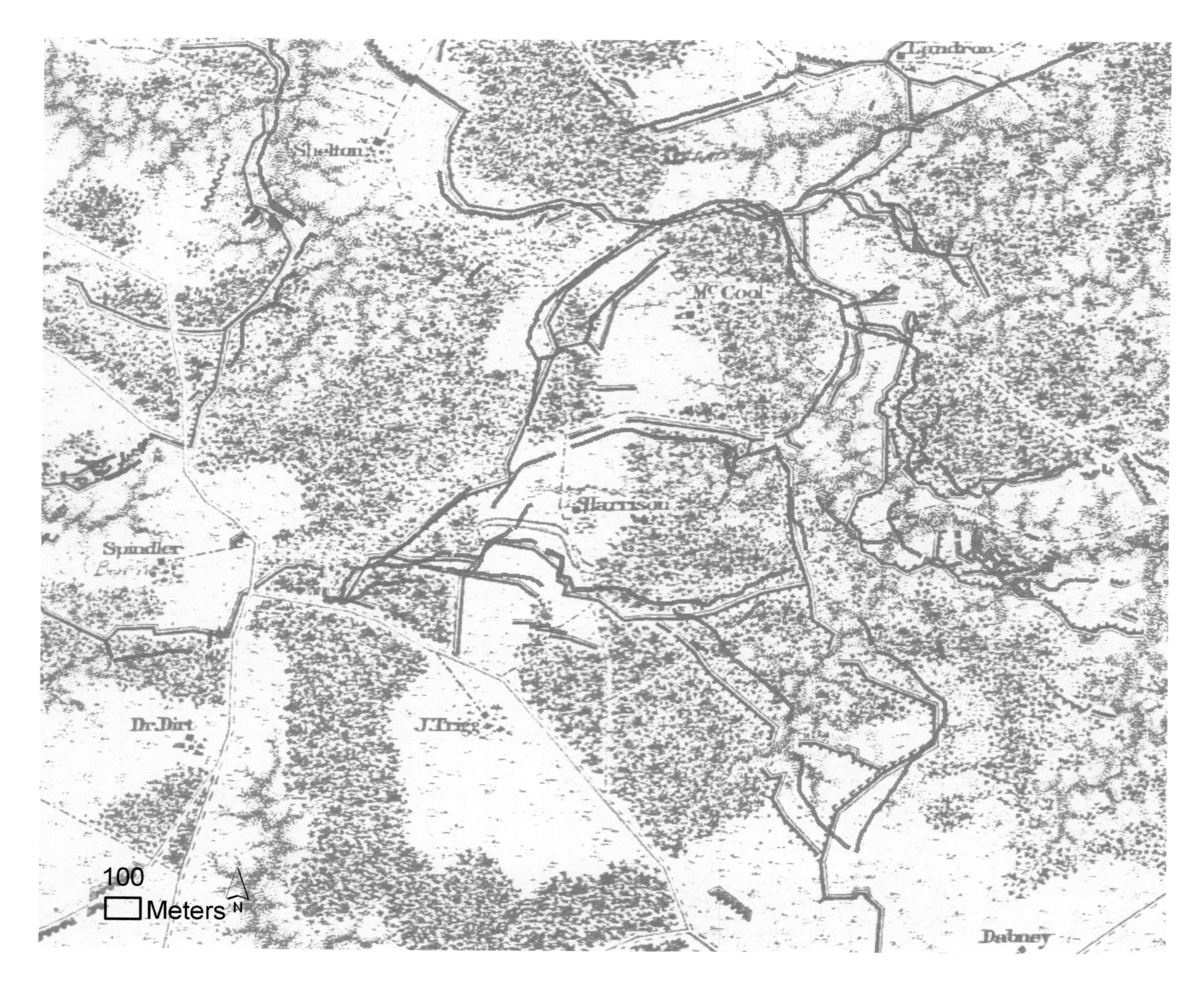

Figure 3. Detail of military map of Spotsylvania Battlefield, showing the area known as the "Bloody Angle"
On May 12, 1864, more than twenty thousand Federal soldiers emerged from the fog to overrun a portion of the Confederate lines here. The dawn attack unleashed the longest, most intense firefight of the Civil War—a brutal fourteen-hour torrent of musketry that literally shredded and felled large trees. Earthworks of both sides that survive today were mapped with GPS and overlain in red on this 1867 map. Most earthworks on the original map, shown as solid brown lines, fall within twenty-five to thirty meters of the actual locations, a tribute to the skill of the topographical engineers. Library of Congress Civil War Map Collection.

of managers, maintenance staff, and rangers. Knowledge of trench locations continued to reside in the gray-matter database of a few dedicated specialists, but when specialists transfer to a different park or retire, they take their knowledge with them. Their replacements start from scratch. We were convinced that a GIS database would provide a way to retain specialist knowledge inside the parks, serving as the park's corporate memory. What specialists learned would be recorded in the computer and continue to accumulate. In the case of Fredericksburg, this has proven true at last. A newly hired GIS specialist has dusted off our GPS positions and digital historic maps to serve as baseline data for what promises to be a fully integrated approach to caring for the park's fortifications. Park managers are using the digitized historic maps to develop a plan for returning areas to their Civil War–era appearance, by planting trees where the maps show woodlots or removing trees where there were open fields.

In 1997, Cultural Resources GIS worked with Petersburg National Battlefield in Virginia, an area where seventeen Civil War battles were fought during the 1864–1865 siege of Petersburg. The national park preserves only a small portion of the historic landscape. The park superintendent sought support from local governments and concerned citizen groups to help preserve the areas that fell outside park boundaries. Because Petersburg was so important during the war, military topographical engineers worked nearly two years to complete a detailed map of the entire area at a scale of eight inches to the mile. This map is among the finest examples of Civil War cartography *(figure 4)*. Applying the methodology developed at Fredericksburg, we collected GPS coordinates for confirmed historic locations in the field to generate a digital map of the historic landscape that embraced nearly 88,000 acres. With the map georeferenced, we were able to digitize the old road network and generate statistics. This work yielded 619 miles of turnpikes, by-roads, and farm lanes. Comparing the old roads with the new, we could determine where roads had been straightened, diverted, or abandoned. This is useful knowledge for the field surveyor who wants to understand the structure and form of a historic landscape.

Our map showed that 110 miles of earthen fortifications were constructed during the siege, a much higher figure than previously estimated. By overlaying modern boundary data, we determined

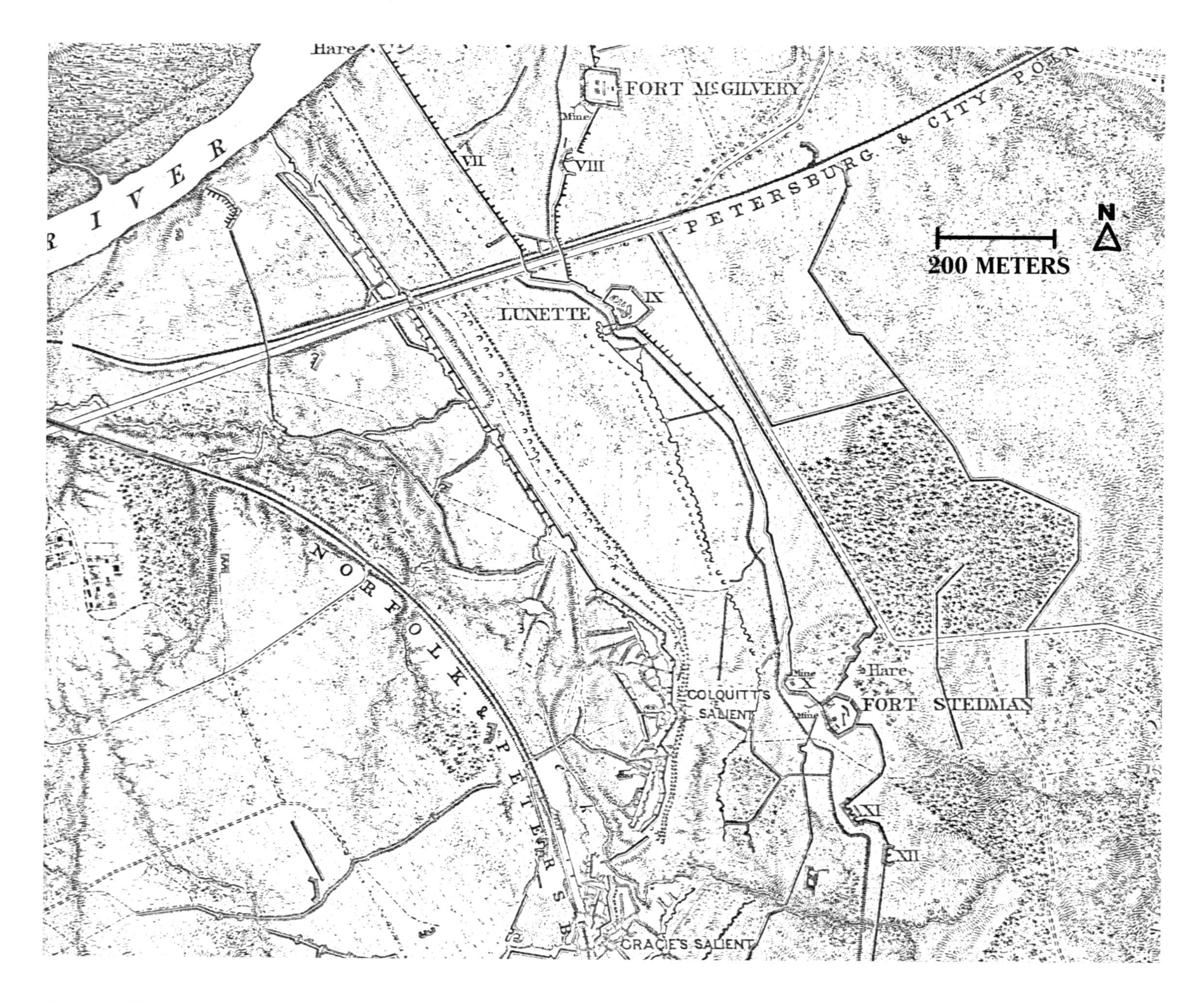

Figure 4. Detail of military map of Petersburg, Virginia
In 1867, military topographical engineers completed this map of Petersburg and vicinity in eight sheets at a scale of eight inches to one mile, making this one of the more detailed mapping efforts of the Civil War. Earthworks were depicted in complex detail. Houses were labeled by the resident's name. The detail shows an area just east of the town of Petersburg and south of Appomattox River where about one hundred yards of no-man's land separated the opposing sides. Map detail courtesy of the National Archives.

that only about 25 percent of the original fortifications were within parklands; the rest were unprotected or destroyed. Following our modern map overlay of historic features into an overgrown area inside the park, a team rediscovered an intact section of Confederate defenses that no one had visited for years. Park personnel used GPS to add these trenches to their database of cultural resources. In the near future, visitors will follow a hiking trail with wayside exhibits to the site to experience more of the story of the Siege of Petersburg.

We then turned to the seventeen types of land use and land cover shown on the historic map, such as tilled fields, woods, orchards, swamps, and town lots, and digitized the different areas as polygons. These generated a computerized picture of the historic landscape *(figure 5)*. The incredible detail of the historic map showed us that the armies felled 6,500 acres of trees to build fortifications and winter huts and to burn as firewood. (The army cartographers depicted these clear-cut areas by drawing in countless tiny stumps.) This amounted to 20 percent of the original woodlands. A historical ecologist could have a field day studying the land-use pattern at this scale, measuring how warfare altered the landscape.

The historic map also showed the locations of 4,882 houses, barns, and other structures in the town of Petersburg and outlying areas. Through various experiments we estimated that most places on the historic map differed from actual location by only forty to fifty yards, close enough to get a surveyor within striking distance of a site. A GPS receiver not only collects spatial coordinates but also allows a surveyor to navigate to any set of coordinates. We selected ten structures from the historic map, entered coordinates, and followed the GPS out to the sites. We found two standing antebellum houses, still occupied, a foundation pit and rubble from one long destroyed, and three modern houses shaded by very old trees. Local residents confirmed that the new houses had replaced older ones and that the farms had been in operation since at least the Civil War. Three other locations yielded factory buildings and a convenience store, suggesting that progress had claimed our prey. The last site was buried in a pine thicket so dense that deer had not forged a path through it. Lacking a machete, we abandoned the search. This exercise convinced us that the overlay coordinates could serve as the basis for a systematic survey of historic structures in the Petersburg area.

Figure 5. Land-use map of Petersburg and vicinity derived from a military map
The historic map of 1867 depicts seventeen different land uses covering an area of about 88,000 acres. Dark green patches were standing forests; light green areas depict the 6,500 acres clear-cut by soldiers to build defenses, clear fields of fire, or burn as firewood; straw-colored patches in the upper left, north of Appomattox River, were croplands beyond range of Federal soldiers. At the time of the Civil War, 4 percent of the landscape was developed for dwellings, 4 percent cultivated, 45 percent in pasture, and 37 percent wooded.

When finished, surveyors could report the number of antebellum structures still standing and how many were gone, valuable information for people who promote preservation.

This time we delivered our data into eager hands. The park's full-time GIS specialist daily churns out maps to illustrate battlefield troop movements, historic landscape features, and aspects of the park's management plan. The park superintendent at last can stand in a county zoning hearing on equal footing with real estate developers, who bring impressive maps and expensive displays showing their vision for the future. In several cases, the superintendent's cartographic evidence helped convince the zoning board that a historic site should be preserved rather than paved.

The battle that raged at Bentonville, North Carolina, in March 1865, just weeks before the surrender at Appomattox, was as fierce as any in the war.

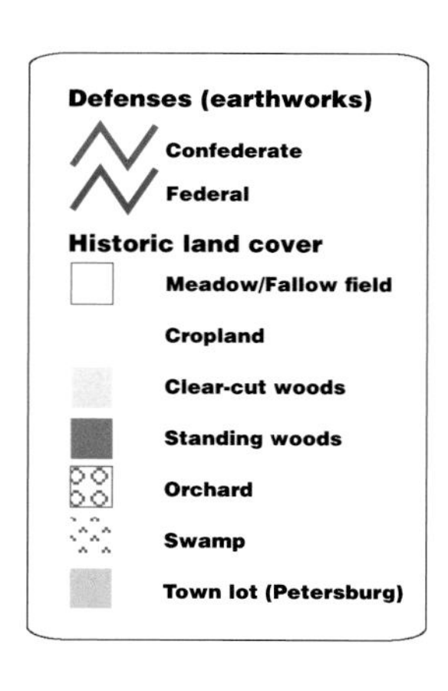

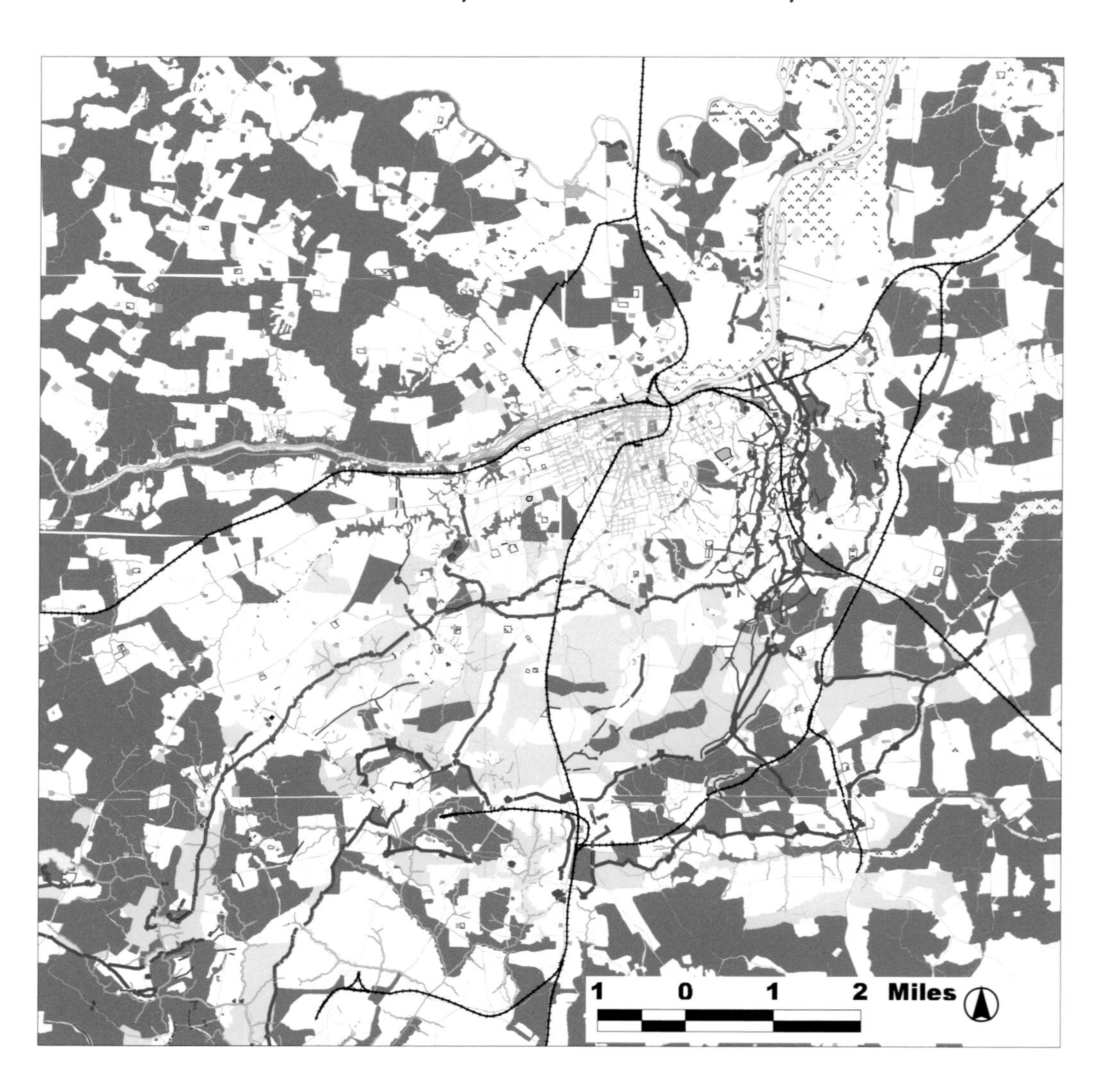

Over three days, the armies of North and South fought trench warfare on a front nearly six miles long. Over the decades, most of the trenches were plowed under for agriculture. What remains is fragmented and often difficult to find in the second-growth forests. Cultural Resources GIS teams assisted Bentonville Battleground State Historic Site in 1999 by mapping the often shallow remains of these trenches, first to build a full inventory of resources, second to see what the inventory might tell us about the battle. The state park owns a tiny portion of the battlefield, but our work was heartily supported by local farmers who allowed access to their land and sometimes even told us where to look.

In the end, we mapped three miles of trench line, all that remains of an estimated fourteen miles of defenses. The inventory helped save about nine hundred feet of earthworks from the ravages of mechanized clear-cutting. The loggers agreed to work around them and left a buffer of standing trees as protection. Our survey also resulted in important revisions of modern battle maps. Park historians had long assumed that the right of the Confederate line made a sharp turn and extended north overlooking a deep ravine. The historic map overlay showed the line descending across the ravine, and we discovered surviving earthworks to prove it. This finding pushes the far right of the position nearly half a mile in front of where it was presumed to be and forces a reexamination of combat on this end of the battlefield. Another trench fragment convinced us that historians, relying on descriptions in battle reports, had misplaced the center of the position by about 200 yards. The evidence of the ground enables park historians to better tell the story of the Battle of Bentonville.

In many cases, the needs of cultural and natural resource preservation intertwine. Stones River National Battlefield in Murfreesboro, Tennessee, harbors most of what remains of Fortress Rosecrans, built in 1863 to defend a Federal supply depot. The fortress, with a perimeter of two and a half miles, originally consisted of a twelve-foot wall of earth fronted by a deep ditch. Attack against it would have been suicidal. Town growth over the years destroyed more than 80 percent of the fortress. To make the surviving earthworks more visible, the park removed most of the trees that had covered the site, but that increased erosion. Biologists established a series of test plots to determine the best mix of native grass species and planting techniques to protect the earthworks from the elements.

In 2001, a team from Cultural Resources GIS assisted the process by locating the test plots and monitoring stations with GPS and making a map that biologists will use to follow the results of their efforts. In addition, we used field measurements to build a three-dimensional model of the earthworks.[3] It helps managers and park visitors visualize the structure of the fortress and grasp its geographical rationale and effectiveness *(figure 6)*.

Battlefield resources include more than the physical remains of battle. In the decades following the Civil War, veterans groups returned to battlefields to erect monuments to their fallen comrades. Today, these monuments provide a way for visitors to commune with the past. Many park visitors have one question: "Where did my ancestor fight?" To answer it, park staff must identify the ancestor's military unit, locate its monuments, markers, and descriptive tablets on a map, and then direct the visitor along unfamiliar roads and trails. This service is made much easier with GIS technology. At Chickamauga National Military Park in northern Georgia, for instance, more than a thousand monuments, markers, and tablets dot the historic landscape. A

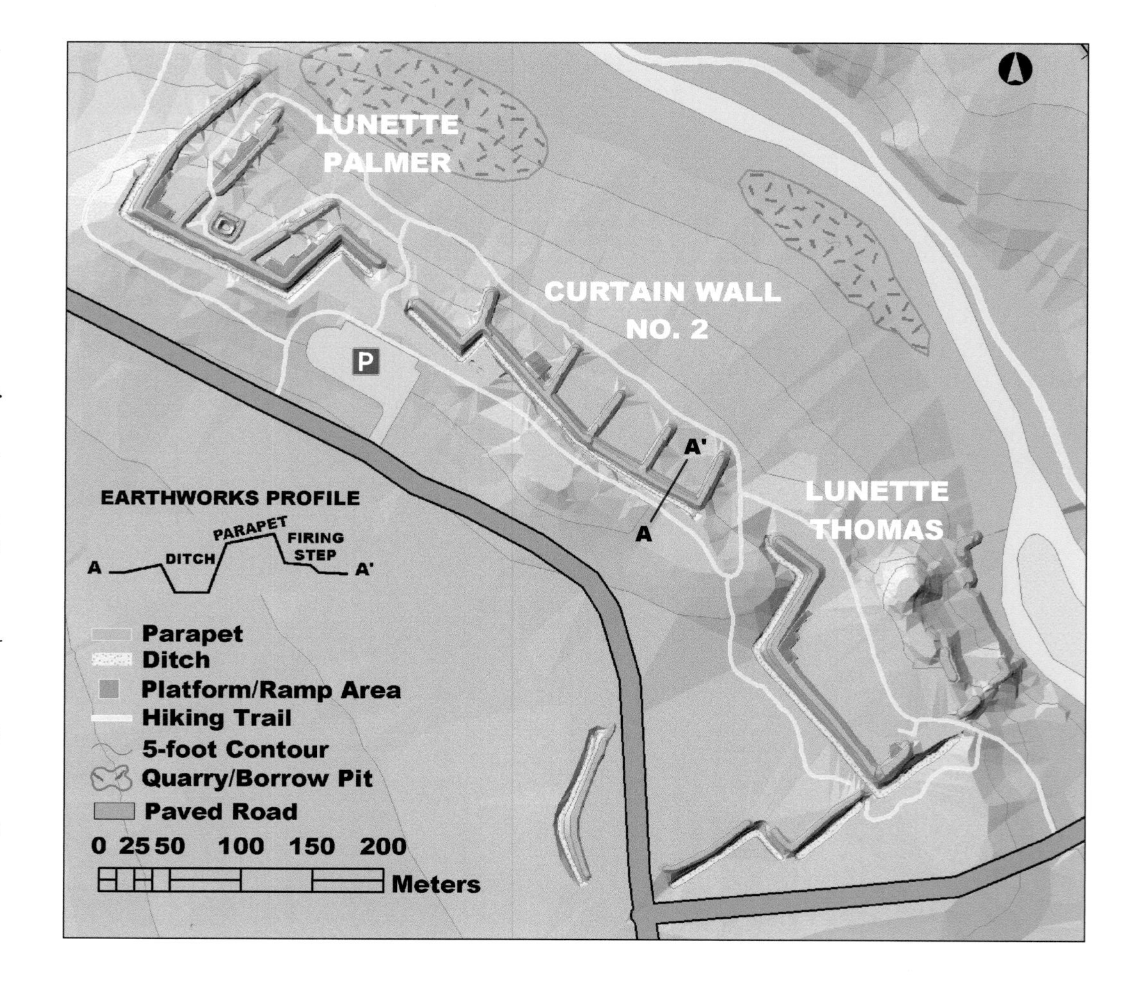

Figure 6. Surviving portion of Fortress Rosecrans, Murfreesboro, Tennessee
Lunettes Palmer and Thomas (structures for artillery) and Curtain Wall No. 2 (the infantry earthwork that connected them) made up the southwest wall of Fortress Rosecrans. The fortress originally formed a perimeter of alternating lunettes and curtain walls that extended about 2.5 miles. The five-foot-deep ditch at the base of the twelve-foot-high parapet served as a dry "moat" in front of the structure. Cannons, placed on platforms built into the angles of the lunettes, could generate a deadly crossfire in front of the earthwork. Short sections of parapet, called traverses, extended to the rear at a right angle to the main line to protect soldiers from incoming artillery rounds. As can be seen in this 3-D representation, much of the right half of Lunette Thomas was hauled off long ago for fill dirt.

park map from the 1930s showed these monuments and markers as clusters, not individual objects. No map showed the locations of the hundreds of tablets. In a little over a week, three teams surveyed all the objects, along with 199 cannons and thirty-five miles of hiking trails. Bringing the data into GIS resulted in the first park map showing all individual elements of the commemorative landscape. Eventually, each monument or marker on the digital map will be linked to a database that provides the full text of the inscription and a photograph. With everything in one place and linked spatially, the park ranger can search by unit, find monuments and markers, even print out a guide map. The ranger will save time; visitors will have a more meaningful experience in the park.

Caretakers of military cemeteries have a similar challenge. Petersburg National Battlefield cares for the Poplar Grove National Cemetery, which contains the graves of more than six thousand Federal soldiers. Hundreds of visitors each year ask how to find an ancestor's grave. The answer will soon be generated by an interactive cemetery grave locator *(figure 7)*.[4] Selecting a name from the burial register highlights the plot on a basemap of the

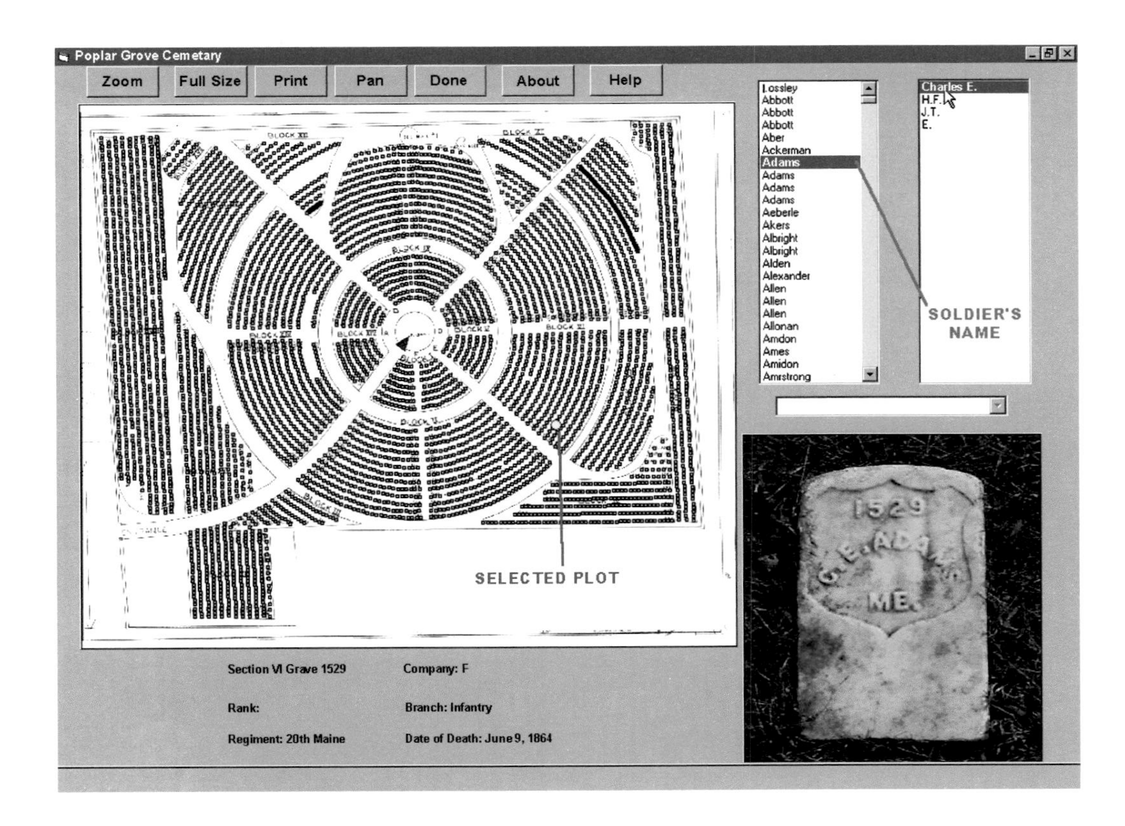

Figure 7. Screen capture of Poplar Grove cemetery grave locator

This MapObjects® application was developed by staff at Cultural Resources GIS to help visitors find the gravesite of an ancestor. The user selects last name, then first name or initials, to highlight the corresponding cemetery plot on the map. This simultaneously displays a photograph of the gravestone and soldier information, which can be printed.

cemetery. A photograph of the gravestone is displayed along with any other information that might be known about the individual, such as rank, unit, and date of death. Visitors will be able to print the information and follow the map to find their ancestor's resting place.

The problems of preserving resources are even greater for sites related to earlier wars. Congress appropriated funds in 2000 to determine which battlefields and other sites of the Revolutionary War and War of 1812 are still intact and which have been lost or fragmented by development. Historians identified 256 battlefields and 527 properties involved in these conflicts, such as forts, arsenals, shipyards, winter encampments like Valley Forge, or places where politicians debated, like Independence Hall. The sites are spread across thirty-one states and the District of Columbia. Working with the National Park Service's American Battlefield Protection Program, Cultural Resources GIS has trained seventy-five surveyors to assess the components of a battlefield landscape, its condition, and the threats to its survival, if indeed it is not already lost to urbanization or suburban sprawl. Each surveyor carries a small GPS unit to plot and describe structures, old roads, fords, forts, and other features. They also define battlefield areas on digital versions of the U.S. Geological Survey quadrangles.[5] This spatial data will be maintained in a central GIS database and linked with descriptions of battles, photographs, and information on preservation efforts. Satellite images can be consulted to see which battlefields are in the path of urban growth, so that planners can anticipate how to preserve the most land for the least cost. Congress will use all this information to set priorities for the limited pool of preservation dollars. When the project is complete, there are plans to offer Internet access to selected information and spatial data through ESRI's ArcIMS® clearinghouse.

These applications reach far beyond the battlefields that have served as our field laboratories. As more historic maps are updated with spatial coordinates, larger expanses of the historic landscape become available for study. As more sites are accurately plotted with GPS and brought into GIS, our fund of knowledge and our nation's corporate memory will grow. Resource management and preservation will become more systematic and effective. Visitors will have more ways to understand and appreciate the past. Spatial technologies are helping the National Park Service tell its stories.

Further reading

American Battlefield Protection Program. *Battlefield Survey Guide.* Washington, D.C.: National Park Service, 2000.

Kennedy, Michael. *The Global Positioning System and GIS: An Introduction.* Chelsea, Mich.: Ann Arbor Press, 1996.

Kennedy, Frances H., ed. *The Civil War Battlefield Guide,* 2d ed. Boston: Houghton-Mifflin, 1998.

Lowe, David W. "Field Fortifications in the Civil War." In *North and South: Special Infantry Tactics* 4, no. 6 (Summer 2001): 58–73.

McElfresh, Earl. *Maps and Mapmakers of the Civil War.* New York: Henry N. Abrams, 1999.

National Archives and Records Administration. *A Guide to Civil War Maps in the National Archives.* Washington, D.C.: National Archives, 1986.

Steede-Terry, Karen. *Integrating GIS and the Global Positioning System.* Redlands, Calif.: ESRI Press, 2000.

Stephenson, Richard W. *Civil War Maps: An Annotated List of Maps and Atlases in the Library of Congress.* Washington, D.C.: Library of Congress, 1989.

Online resources
American Battlefield Protection Program home page: www2.cr.nps.gov/abpp (*Battlefield Survey Guide* and related items)

Cultural Resources GIS home page: www2.cr.nps.gov/gis (information on the field uses of GPS and GIS, survey reports, and examples of battlefield mapping projects)

Notes

1. GPS was developed by the U.S. Department of Defense and is one of the technologies underlying the American military's ability to deliver "smart bombs" on target. Twenty-four orbiting satellites broadcast time-coded radio signals that are received on the ground. Signals from four satellites are processed in a handheld computer, which calculates distances by comparing the time the signal left the satellite to when it arrived. This enables it to triangulate the receiver's ground location. Trimble™, one of the large GPS companies, describes this process in detail at www.trimble.com.
2. To georeference the map, we used the tools in ArcInfo, ESRI's high-end GIS system.
3. We built the 3-D model using ArcView software's ArcView Spatial Analyst extension.
4. One of our specialists created the locator as an ESRI® MapObjects application.
5. The USGS offers georeferenced map images for most quadrangle maps on the Internet in digital raster graphic (DRG) format. Surveyors use a MapObjects application, developed by staff of Cultural Resources GIS, to draw areas on-screen, with the DRG as background. This allows us to import their findings directly into GIS.

CHAPTER FIVE

IMMIGRATION, ETHNICITY, AND RACE IN METROPOLITAN NEW YORK, 1900–2000

Andrew A. Beveridge

In 1910, two-fifths of New York City's 4.8 million people came from abroad, making it probably the most cosmopolitan city in the world. Over the next ninety years, the city underwent many changes in territory, population, and racial and ethnic composition. The five boroughs of New York City, consolidated in 1898, became the center of a metropolitan region that now extends into four states and includes more than thirty counties. People living in the urban core of this area nearly doubled to almost nine million, while the metropolitan area as a whole grew to more than twenty-one million. The European diversity of 1910 became truly global diversity by the year 2000. Diversity, however, has a different geography now, with many more national and ethnic groups represented in a giant metropolitan area that is even more highly segregated than the city was at the beginning of the twentieth century.

It is difficult to document and to comprehend the complex patterns of immigration into and out of New York. Many neighborhoods have been repeatedly reconstituted, changing language, religion, complexion, and economic status. This chapter focuses on broad changes that have altered the basic geography of native-born and immigrant, white and nonwhite, rich and poor over the last century. We pay particular attention to the urban boom of 1905–1940 and the postwar explosion of suburbia, both of which reconfigured New York's population in fundamental ways. We also show how the rapid waves of inward and outward migration made the city more diverse even as it crystallized patterns of segregation. All of this is made possible by using a GIS with appropriate map and census data.

Over the decades, social scientists, planners, and public officials trying to understand these changes have turned to the U.S. Population Census. Their desire to understand New York City influenced how the census presented data about urban areas, for the "Census Tract Movement" began in New York in 1906, when urban planner Walter Laidlaw suggested that the city be divided into units according to population. In 1910, New York was one of the first eight American cities to be divided into census tracts. Each tract in the most populous parts of New York averaged about eight city blocks; tracts were larger elsewhere. The tract system was gradually applied to other cities. Washington, D.C., was added in 1920 and six more cities in 1930. By 1990, the Census Bureau divided the entire United States into tracts.

We began our study of New York City's demographic history by constructing a GIS that links social and economic data to census tract boundaries from 1910 to 2000. The GIS also includes information on subways, bridges, tunnels, and main thoroughfares. We then used the GIS to examine how the influx and outflow of population has altered the city: the shifting location of racial and ethnic clusters; the degrees of segregation and diversity; the density of population; and the relationship of all these to economic inequality.

Although census tract boundaries have changed with time, by using GIS technology it is possible to study population change over a period of ninety years. It is extremely important to have the tract map for each census aligned to a common basemap, as Ian N. Gregory and Humphrey R. Southall explain in chapter 9. Doing this requires creating tract boundary files for each census, which for New York City required months of work. Beginning with the 1990 tract data and tract and block map boundary files provided by the Census Bureau, my associates and I used scanned maps, paper maps, and Census Bureau tract comparability materials to edit 1990 tracts to become 1980, 1980 to become 1970, 1970 to become 1960, and so forth. Recently, we integrated the 2000 maps and data into the GIS. Having all these years of census data in a common form allows the assessment of changing settlement patterns and segregation in a rigorous manner, and makes it possible to use standard methods of population allocation when boundaries shift. Once available, other researchers can use the base GIS data to add material on cultural institutions, events such as riots, land use, and other features to investigate other phenomena. Here we concentrate on general settlement patterns.[1]

A new project has begun that holds the promise of creating similar data and maps for the United States as a whole. The National Historical Geographic Information System, based at the Minnesota Population Center at the University of Minnesota and directed by John Adams, will digitize all available tract data and boundaries for 1910–2000 and all available county-level data and boundaries for 1790–2000. The data and boundaries will be made available publicly over the Internet, making analyses such as those we have done for New York City possible for almost any locale in the United States. Our New York City maps will become integrated with this project, and my associates and I have uncovered data and maps for many of the largest cities in the United States going back to early decades of the twentieth century.[2]

In 1910, the island of Manhattan was home to about 2.3 million people. One-third lived south of Eighteenth Street *(figure 1)*. Nearly half of Manhattan's population was foreign-born, and the Lower East Side was the most heavily immigrant neighborhood at that time. More than one-third of its 600,000 residents were Jews from Russia, Hungary, Romania, Galicia, and the Levant. Another 200,000 were immigrants from Ireland, Italy, or other European countries.

About 53,000 Italians lived on the Lower West Side, a neighborhood of

Figure 1. Population density and immigrant neighborhoods in New York City, 1910
East Harlem, the Lower East Side, and the Lower West Side had the highest population density in turn-of-the-century New York. The density of immigrants in the Lower East Side was most remarkable, particularly considering that most lived in tenements no more than four stories high.

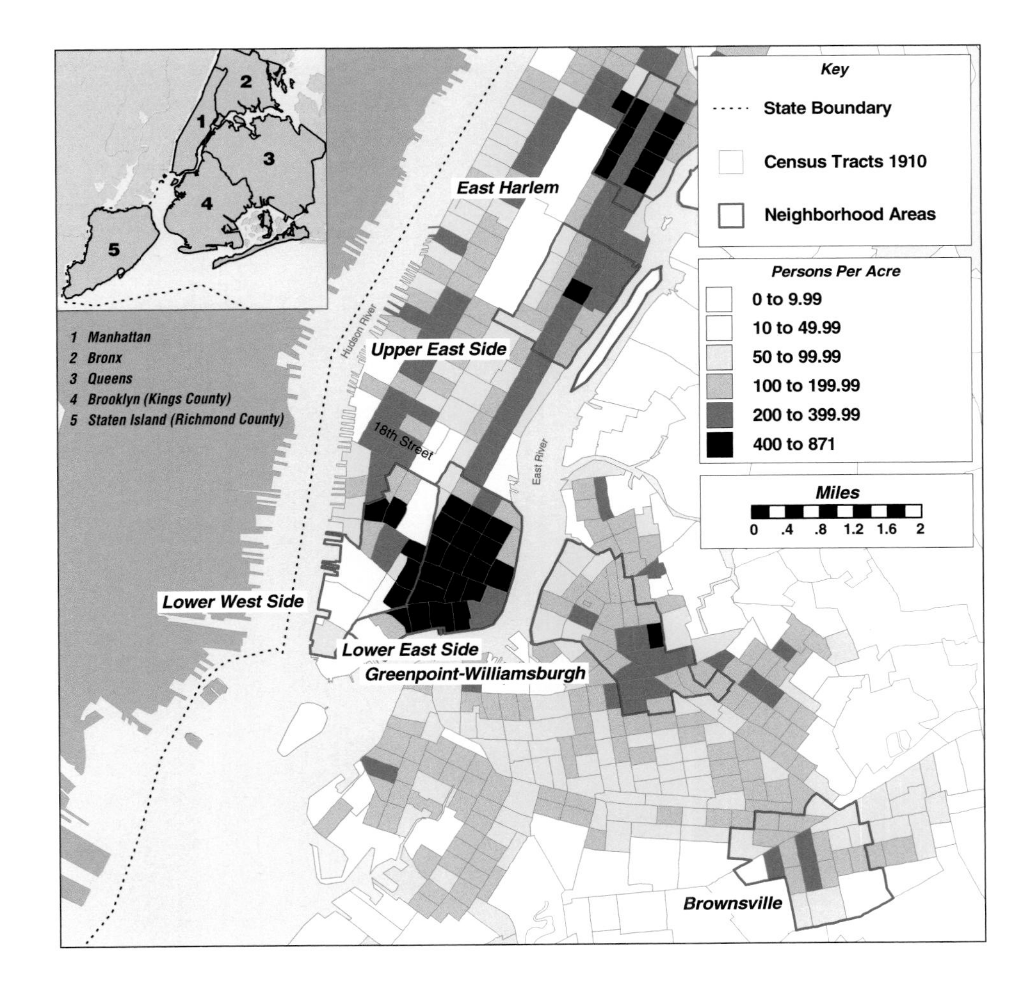

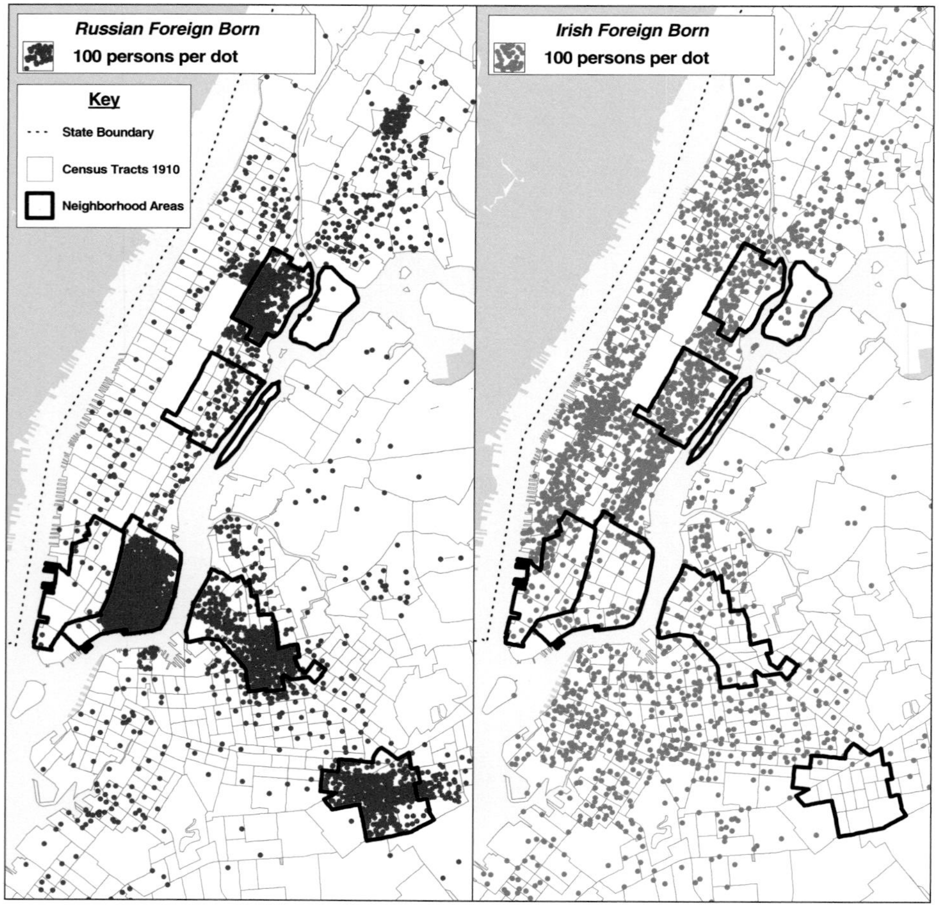

about 185,000 people within walking distance of factories near the Hudson River. Many Neapolitans and Calabrians lived in "the Bend," a district along Mulberry Street. Genoese were prominent along Baxter Street, and several hundred families from the Sicilian fishing town of Sciacca lived along a few blocks of Elizabeth Street. A second "Little Italy" at that time was up in East Harlem; its immigrant population came mainly from southern Italy. The Greenpoint-Williamsburgh neighborhood, across the East River in Brooklyn, developed as an immigrant settlement for Russians, Italians, Germans, Austrians, and Irish after the Williamsburgh Bridge opened in 1903. By 1920, the area contained about 360,000 people. Population density was greatest on the Lower East Side, where it averaged more than 300 persons per acre. Some tracts were as dense as 871 persons per acre.

Lower Manhattan's tenement districts became synonymous with big-city slums, largely due to the pioneer muckraking journalism of Danish immigrant Jacob Riis. His reports and photographs of tenement life shocked the nation. "Tenement houses were designed for twenty families," he wrote, "but these buildings often housed as many as one hundred residents, as well as their boarders. Upwards of 200 licensed 'flophouses' with nearly eighteen thousand bunks lined the Bowery and Oliver, Chatham, and Mulberry streets."[3] One of the reasons Laidlaw pressed the Census Bureau for data on smaller areas was to be able to document living conditions within these overcrowded districts. His research reported the high densities in the Lower East Side and commented on the lifestyles, health, and education of immigrants.[4]

At the same time, New York's thronging immigrant neighborhoods supported many cultural and civic institutions. Jewish cultural life included Yiddish

Figure 2. Russian and Irish settlement patterns, 1910
The density of Russian immigrants in East Harlem, the Lower East Side, Greenpoint-Williamsburgh, and Brownsville reflects the pull of chain migration, as Jewish immigrants took up residence and work near relatives and friends. Irish immigrants' more scattered pattern was the result of many factors, including their longer residence in the city.

theaters, temples, schools, and clubs promoting unionism and radical politics. Religious festivals marked Catholic holy days in the streets of New York. In lower Manhattan, one could find Russian Orthodox churches, Jewish synagogues, and Italian social clubs all within a few blocks of one another.

The patterns of immigrant settlement varied between groups *(figure 2)*. For instance, many of the Russian immigrants who came to New York City between 1880 and 1913 settled in one of four neighborhoods. One of the Russian enclaves, the Brownsville neighborhood in Brooklyn, became the city's largest Jewish slum, with sweatshops and pushcarts and no sewers or paved streets. A Jewish entrepreneur bought land there in 1887, built tenements, and enticed several Jewish garment makers to relocate from the Lower East Side. The opening of an elevated train line in 1889 contributed to Brownsville's growth. It was home to 81,000 immigrants in 1910, almost three-quarters of them Russian Jews who worked in nearby factories. Irish settlement shows a very different pattern. Large numbers of Irish immigrants came to New York during the Great Famine in the middle of the nineteenth century. As Irish immigration continued, newcomers, long-standing residents, and second- and third-generation Irish took up residence throughout the city. By 1910 almost every census tract in Manhattan, as well as many outlying districts, had some Irish population. British immigrants and Germans were similarly dispersed.

Almost as soon as subway construction began in 1904, the new lines of transportation began to influence the city's residential pattern *(figure 3)*. The subway initially spurred the development of upper-middle-class neighborhoods in upper Manhattan, Brooklyn Heights, and exclusive areas of the Bronx. As subway lines extended deeper into Queens, Brooklyn, and the

Figure 3. Subway construction and population growth
The map on the left shows the relationship of residence and public transportation in 1905, shortly after the first subway lines were constructed. By 1940, most of New York's subway lines were complete. The subways were instrumental in shifting population from Manhattan and northern Brooklyn into southern Brooklyn, the Bronx, and Queens.

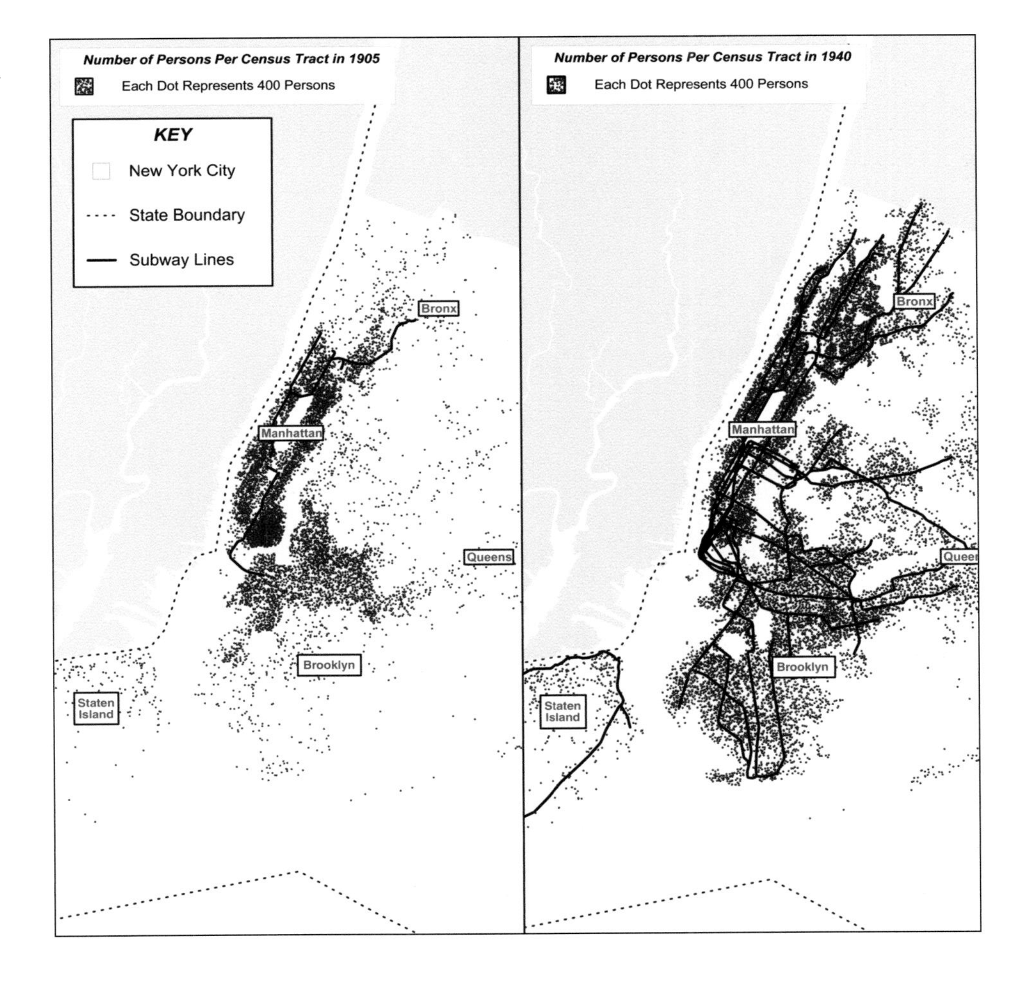

Figure 4. Population shift in New York City, 1905–1940
This map locates the changes of population suggested in the previous figure. Tenement districts on the east and west sides of Manhattan and in Greenpoint-Williamsburgh, in northern Brooklyn, became much less densely populated as new, less dense housing developments several miles outside the city center attracted young couples and new immigrants.

Bronx, about 250,000 people left the Lower East Side, many of them moving from tenements to two-family homes in newly platted neighborhoods such as Woodside in Queens and Bensonhurst and Bayridge in Brooklyn. New immigrants joined them there. The growth of Queens was particularly dramatic. It went from 280,000 people in 1910 to 467,000 in 1920 and 1.29 million in 1940.

Figure 4 maps the growth and decline of population across New York City from 1905 to 1940. While densely settled immigrant neighborhoods lost large fractions of their populations, most of the rest of New York City grew very rapidly. In addition to opening up new areas for residential development, the subways promoted a new pattern of relatively low-density housing that consumed large areas of land. Single-family houses, duplexes, and low-rise apartment complexes dotted the outer boroughs and filled the rest of Manhattan. The sort of city life described fondly in E. B. White's 1948 essay, "Here is New York," and chronicled on television by Jackie Gleason's situation comedy, "The Honeymooners," replaced tenement life. By the beginning of World War II, New York City had achieved the physical scale and density that we know today. The city's street grid was established, as were most of its parks.

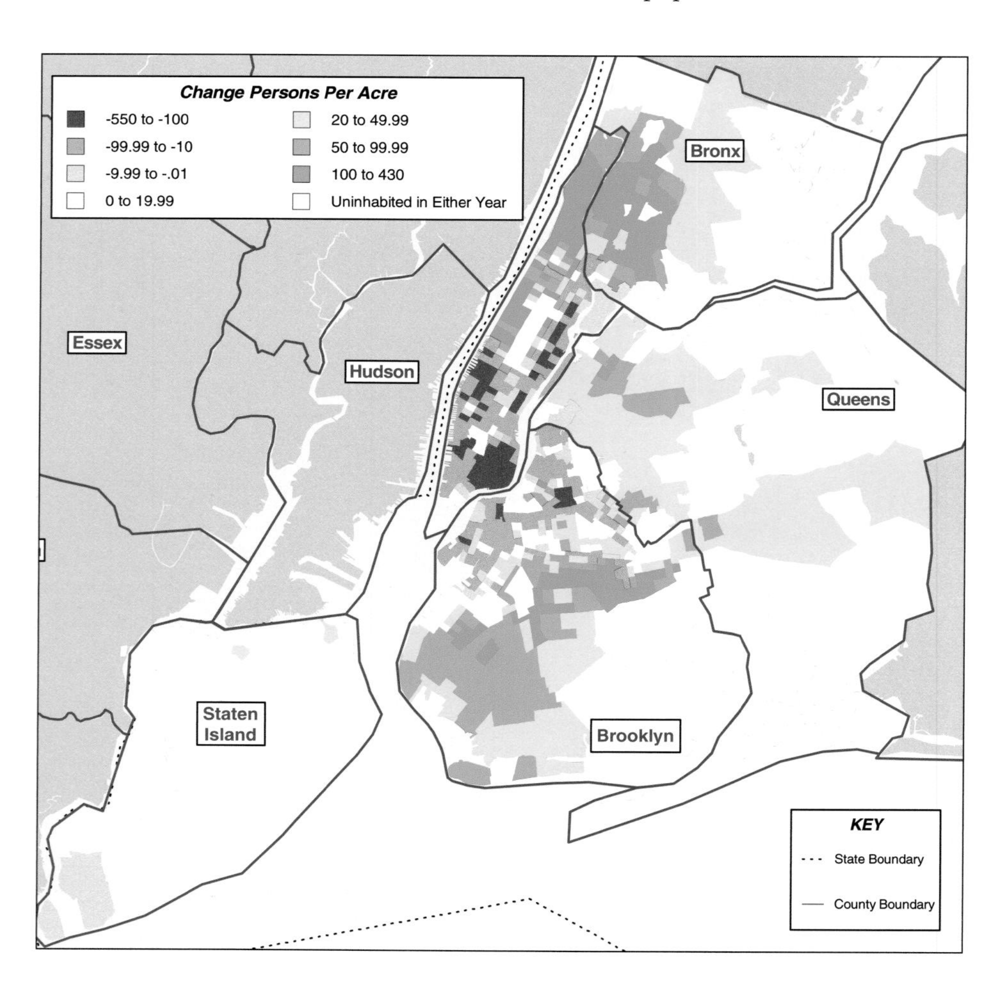

After 1940, New York burst the bounds of the five boroughs to become a true metropolis. The precise limits of the New York metropolitan area have been a subject of debate ever since the Census Bureau declared its existence in 1940.[5] The Regional Planning Association, for instance, defines the New York region as thirty-one counties. Others say it includes everything within a seventy-five-mile radius of Times Square. We follow the Census Bureau's current definition, namely the twenty-nine counties that were completely contained in the New York Consolidated Metropolitan

Statistical Area starting in 1995. The Census Bureau uses commuting, transportation, and communication patterns to designate metropolitan areas. So-called designated market areas (DMAs) follow very similar geographies.

We classified the counties into three different groups: urban core, near suburbs, and periphery *(figure 5)*. The urban core consists of all the boroughs of New York except Staten Island, as well as Hudson and Essex counties in New Jersey. We excluded Staten Island because even today it is not entirely urbanized, and is more like a suburb than the rest of New York City. Hudson County includes Jersey City, Hoboken, and Bayonne, while Essex County includes Newark, one of the most concentrated urban environments in the country. The counties designated as near suburbs include Staten Island, Nassau, and Westchester in New York, and Bergen, Passaic, Morris, and Union in New Jersey. Each of these counties is adjacent to a core county. The seventeen counties in the periphery are areas of rapid growth, including counties with so-called "edge cities" like Stamford, Connecticut, and the area near Route 1 around Princeton, New Jersey. Admittedly, classification by county is somewhat crude. However, any differences would be even more striking with a more nuanced classification of the metropolitan area.

In 1940, the New York metropolitan area had a population of roughly 13 million. Approximately 8.8 million, about 65 percent, lived in the core. Sixty years later, the metropolitan population had grown to 21 million, but the core's 9.0 million people accounted for just 42 percent of metropolitan population. Figure 6 charts the relative decline of urban core population and the sustained growth in the near suburbs and periphery. Within the urban core, waterfront communities grew, as did already urbanized districts in Queens and on Staten Island. Most other parts of the

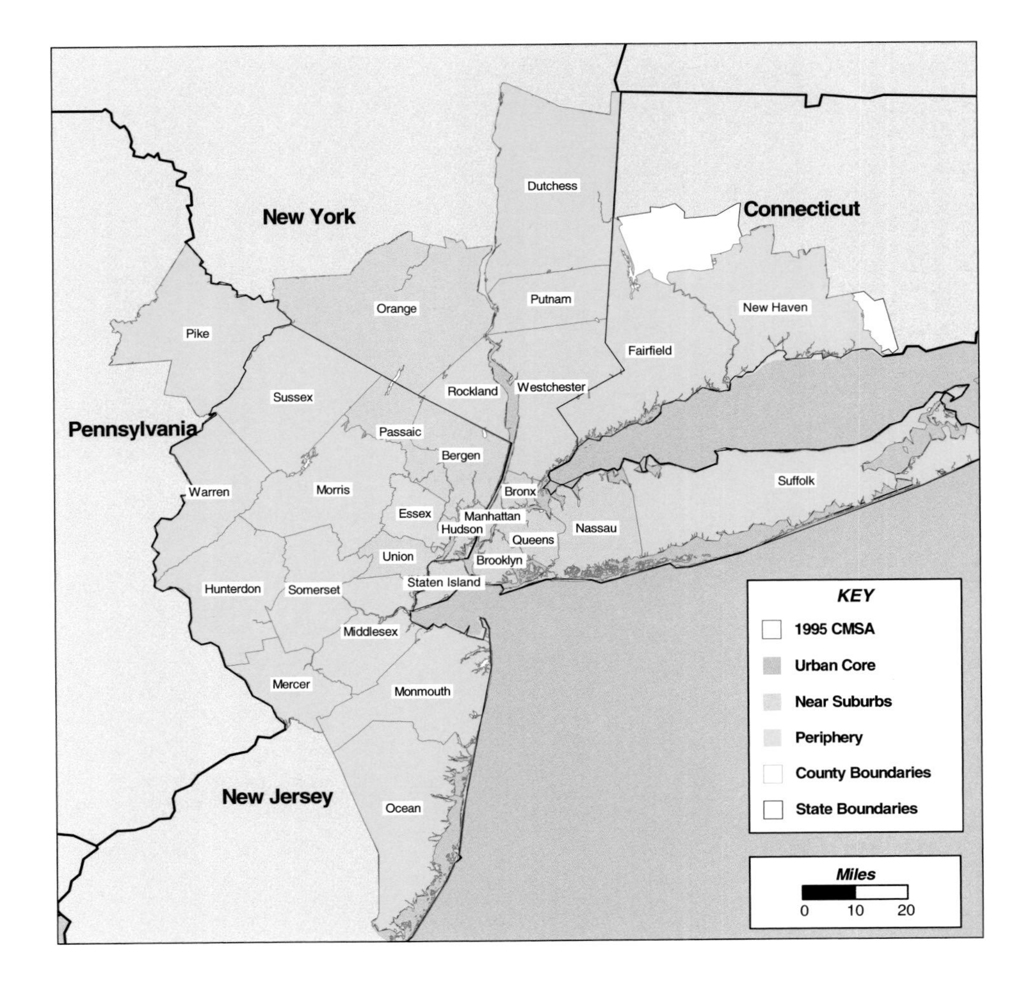

Figure 5. The New York metropolitan area

After World War II, New York's explosive growth blurred the line between the city and its hinterland. The New York metropolitan area is both a statistical entity created by the Census Bureau and an urban reality. This map shows the counties used in the census's definition of Consolidated Metropolitan Statistical area, and the approximately concentric zones of the urban core, near suburbs, and periphery used in our analysis.

Figure 6. Changes in population and median family income
The top chart tracks population change in the urban core, near suburbs, and periphery between 1940 and 2000. Core population dropped after 1970 but climbed steadily after 1980, as gentrification reached lower Manhattan. Suburban population growth, however, was strong and steady throughout the postwar period. The second chart shows the growing income disparities between core residents and suburban New Yorkers.

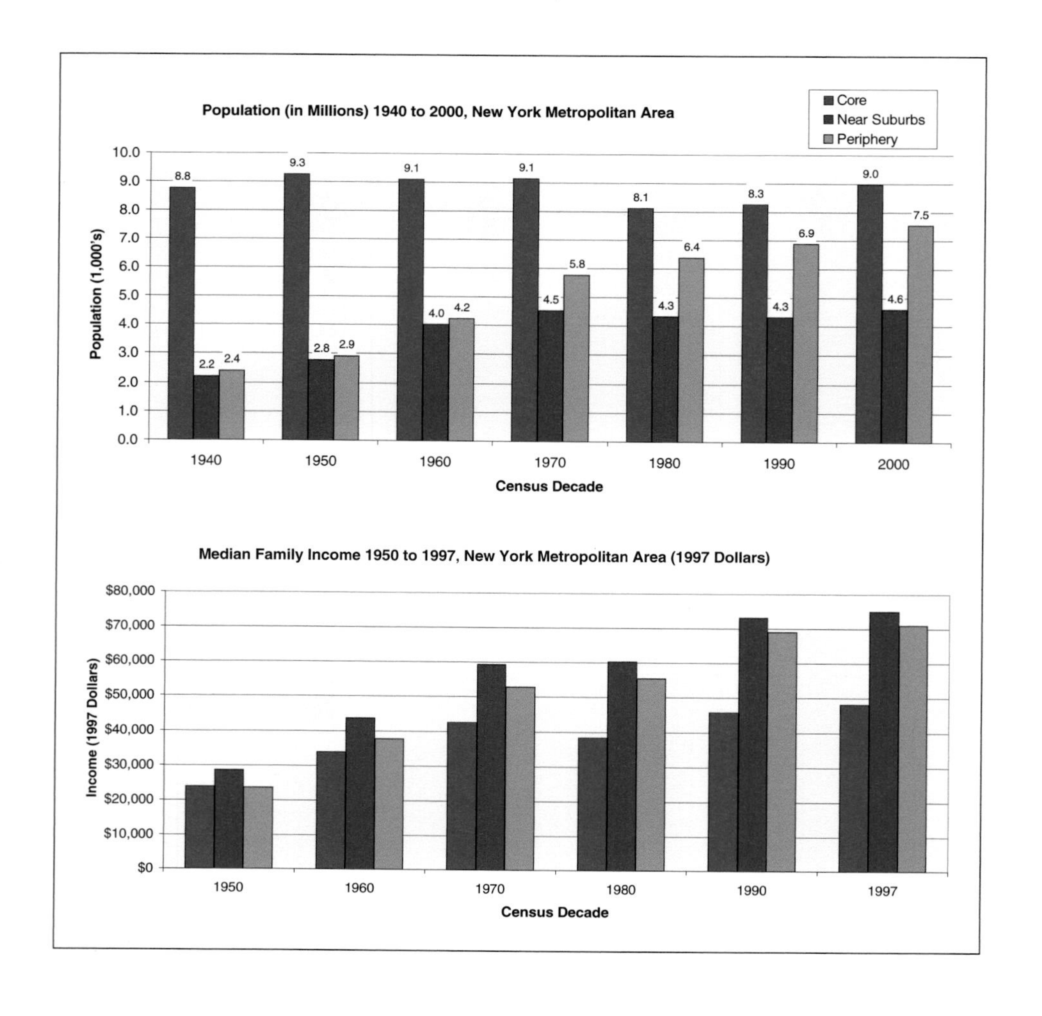

city lost population. Outward migration intensified and moved past the four core boroughs, first into the near suburbs and then to the more distant suburbs in the periphery. In 2000 the total population of Manhattan was just 1.4 million, almost one million fewer people than had occupied the island's crowded neighborhoods in 1910.

Median income follows a similar striking pattern, also shown in figure 6. In 1950, the median family income of $23,794 for the urban core was $207 higher than that of the periphery, but $4,307 (17 percent) less than median income in the near suburbs. In 1997, the latest year for which we have figures, the median income of families living in the core was $48,226, a rise of 103 percent. The near suburbs recorded a gain of 163 percent, to $75,046, while the periphery's median rose 201 percent, to $71,015. Many of those earning large incomes were no longer living in New York City, having fled to the suburbs. This is not to say that New York City no longer has areas of very high wealth, but simply that the "middle New Yorker household" has a much lower income than that of the suburbanite. Indeed, the Upper East Side of Manhattan—along with Smoke Rise, New Jersey; Hewlett Bay Harbor, New York; and New Canaan, Connecticut—had among the highest incomes in the metropolitan area. It is also true that the top fifth of Manhattanites earn on average more than thirty times what those in the bottom fifth earn.

Racial and ethnic patterns also changed *(figure 7)*. All parts of the metropolitan area gained immigrants from other parts of the United States and from abroad. It is difficult to make precise comparisons from decade to decade, because census categories changed. Changes in the general proportions of white (non-Hispanic white) and minority population, however, are very clear. In 1940, New York's nonwhite population constituted about 8 percent of the urban core and about 4 percent in the rest of the metropolitan area. In 2000, the core population was 56 percent nonwhite, compared to 20 percent nonwhite in the near suburbs and periphery. That change initially occurred as black Americans migrated from the South to New York, as to other northern cities, in the 1940s, 1950s, and 1960s. Urban enclaves established in the 1930s by black and Hispanic immigrants from the Caribbean and Central and South America expanded in the 1950s and 1960s and continue to grow today. Nonwhite population also increased in near and peripheral suburbs but at much slower rates. Not only is the urban core much more "nonwhite" now than in 1940, but most of metropolitan New York is more segregated. Commonly used measures show that New York is one of the most highly segregated metropolitan areas in the United States.

Until 1965, immigration quotas were based on country of origin. This favored immigrants from countries that had traditionally supplied much of America's population. European immigration slowed during the Great Depression and then picked up a little right after World War II, but as Europe recovered, it slowed once again. The foreign-born population declined from about 28 percent in 1940 to about 13 percent in 1970. In 1965, immigration law changed so that quotas were based upon an entire hemisphere rather than particular countries. This made more people eligible to come to the United States. Over the last three decades, migration from abroad has once again transformed New York, particularly in the urban core. The percent of foreign-born in the core, roughly 28 percent in 1940, is more than 40 percent today. The proportion of foreign-born has not risen nearly as much in the near and peripheral counties.

Figure 7. Minority population in the metropolitan area, 2000
Black and Hispanic New Yorkers are concentrated in the urban core, as they are in the central-city neighborhoods of Newark, New Haven, and other cities within the metropolitan area. New York's non-Hispanic white population (the "nonminority") dominates the suburbs from Nassau, Westchester, and Bergen counties to the outer ring of commuter suburbs.

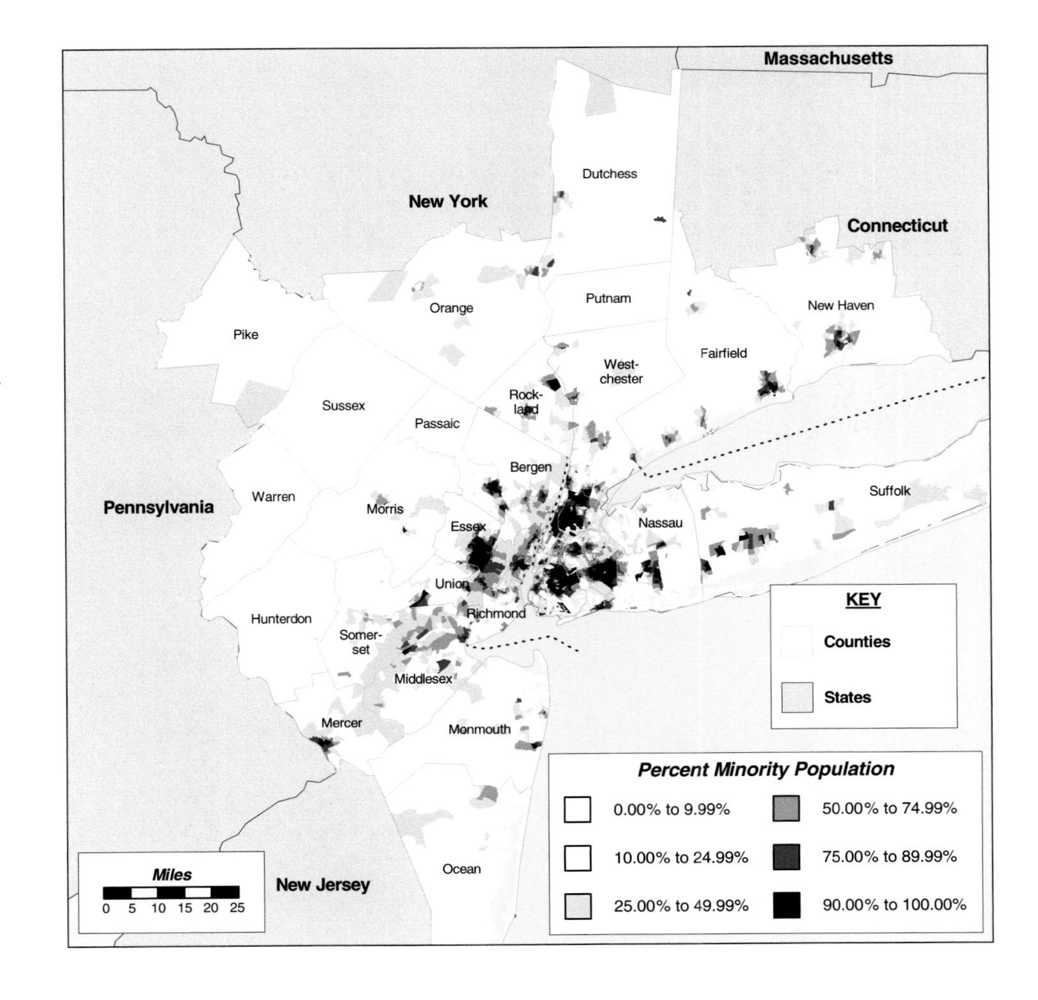

Figure 8. Immigrant enclaves, 1990
Our GIS analysis of the census shows that immigrant neighborhoods are becoming more ethnically homogeneous than they were at the beginning of the last century, the period when we commonly think of New York as the immigrant city. This map picks out six areas with especially concentrated ethnic neighborhoods as recorded in the 1990 census. The exception to this pattern is Western Queens, a remarkably diverse neighborhood with a high proportion of immigrant residents.

Recent immigrants are also creating more intensely concentrated ethnic neighborhoods than existed for any immigrant group in 1910. Figure 8 identifies six highly concentrated immigrant areas in 1990. In 1990, Chinatown was 65 percent Chinese and Washington Heights was 60 percent Dominican, although Dominicans can now be found in many parts of New York City due to continued immigration. Three-quarters of the immigrant residents of Flatbush came from the West Indies in 1990. Results from early releases of the 2000 census, which report detailed racial and Hispanic categories, suggest that ethnic concentrations in these neighborhoods are continuing to rise.

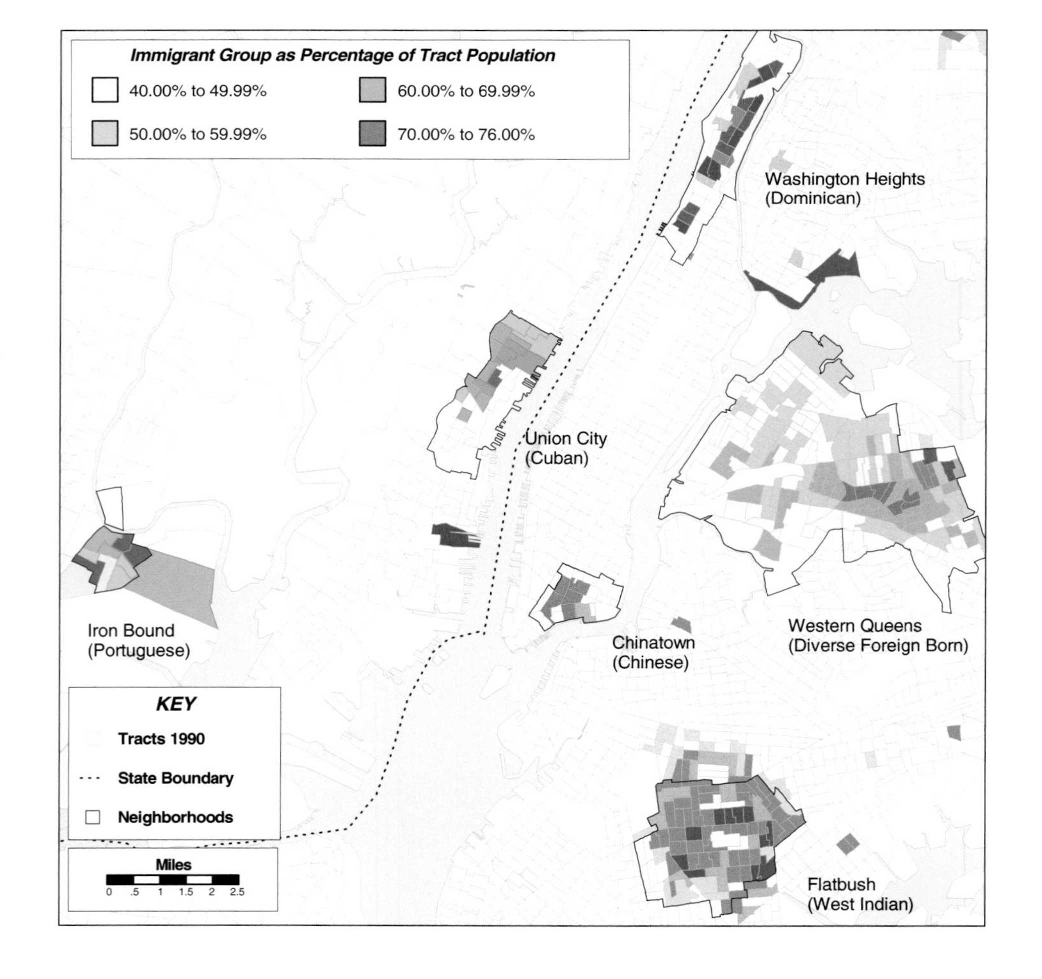

The two enclaves in New Jersey show similar patterns. Almost 60 percent of the more than 18,000 immigrants in Newark's Iron Bound neighborhood came from Portugal. (The name "Iron Bound" comes from the fact that railroad lines surround the area.) Many Portuguese fled the collapse of Portugal's African colonies and the overthrow of the Salazar regime in 1974. Union City, New Jersey, is home to about 60,000 immigrants, including 20 percent of the Cuban Americans in the New York City metropolitan area. In comparison to these neighborhoods, Western Queens is exceptionally diverse. It developed in the 1930s and 1940s as a neighborhood of immigrants from Germany, Ireland, Italy, and Greece. So many nationalities now ride the neighborhood's subway line, the Number 7 IRT, often dubbed "the International Express," that in 1995, President Clinton formally named it one of the sixteen "National Millennium Trails."[6] Many areas of immigrant concentration are reasonably affluent. Indeed, in Queens immigrant households in general have higher incomes than those enjoyed by non-immigrant households, or even those who have recently moved away.

Across the city, our analysis of the census data shows an intensification of the old process called chain migration, in which new immigrants settle among their established countrymen and women, forming ethnic enclaves. We find little evidence of intergroup settlement among post-1970 immigrants except among West Indians. The various groups of immigrants from Asian and Hispanic countries are nearly as segregated from one another as they are from native-born whites, blacks, and European immigrants. The historical GIS revealed that all levels of segregation are higher than they were among immigrant groups in 1910.[7] While further analysis may tell us more about why segregation has increased, at this point we can at least observe that new immigrants are more racially and culturally diverse than former generations. Although Latino immigrants share a common language, Asians do not. At the turn of the twentieth century, immigrants were moving into newly developed, densely packed tenement neighborhoods. Housing was scarce, so immigrants took what they could get. Now established immigrants have more sophisticated channels of communication to guide family and friends to housing and jobs. Apartments, as well as other resources, move swiftly through immigrant networks.

Though segregation is sometimes a matter of choice, it is often a matter of discrimination. Whatever the cause in particular neighborhoods, there are social costs to the growing gulf between New York City and the rest of the metropolitan area. While some suburbs have significant African-American, Latino, or Asian populations, the great majority of people of color live in the central city, segregated by race and class. Many work lower-wage jobs and do not have access to the better schools and job opportunities available in the suburbs. When Jacob Riis wrote *How the Other Half Lives* in 1890, the so-called "better half" walked the same streets as the rest of the population. Now, wealthy New Yorkers live far away in elite, overwhelmingly white suburbs like Chappaqua, New York, and Saddle River, New Jersey. If they come into New York City at all, it is by rail or road, many to work in high-paid jobs. When they go home in the evening, they leave the other half behind in Newark, Harlem, Washington Heights, and the South Bronx. Economic disparity gives the sharp edge to segregation, and poverty makes enclaves into ghettoes.

Acknowledgments

For their assistance in creating the New York City GIS, Andrew Beveridge would like to thank Kenneth "Nick" Trippel, James Schumm, Iris Schweitzer, Michiyo Yamashiki, Susan Weber, Justin Stoger, and Ahmed Lacevic.

The New York Times, Newspaper Division, and the National Science Foundation Course, Curriculum, and Laboratory Improvement Program, Educational Materials Development (grants 9950369 and 008870) provided funding. ESRI provided the software through a grant. We gratefully acknowledge the help of Terry Schwadron, Dylan McClain, Archie Tse, and William McNulty at The New York Times, Myles Boylan at NSF, and Ann Johnson and Michael Phoenix at ESRI. The Web server, workstation, and browser version of the Exploration Software is written in MapObjects with Java™ and C++. The Web client is written in Java. The maps were developed in Atlas GIS™.

Further reading

Cudahy, Brian. *Under the Sidewalks of New York: The Story of the Greatest Subway System in the World.* Bronx, New York: Fordham University Press, 1995.

Foner, Nancy. *From Ellis Island to JFK: New York's Two Great Waves of Immigration.* New Haven, Conn.: Yale University Press, 2000.

Halle, David, ed. *New York and Los Angeles: Politics, Society, and Culture.* Chicago: University of Chicago Press, in press, 2002.

Jackson, Kenneth T. *Crabgrass Frontier: The Suburbanization of the United States.* New York: Oxford University Press, 1987.

Jackson, Kenneth T., ed. *The Encyclopedia of New York City.* New Haven, Conn.: Yale University Press, 1995.

Massey, Douglas, and Nancy Denton. *American Apartheid: Segregation and the Making of the Underclass.* Cambridge, Mass.: Harvard University Press, 1994.

Min, Pyong Gap, ed. *Classical and Contemporary Mass Migration Periods: Similarities and Differences.* Walnut Creek, Calif.: Alta Mira Press, in press, 2002.

United Nations. MapScan software and documentation. New York: United Nations Statistics Divisions, 1999.

United States Census Bureau. TIGER (Topologically Integrated Geographic Encoding and Referencing) System. Washington, D.C.: U.S. Census Bureau, 2001. Available on the Census Bureau's Web site. Go to www.census.gov and check Geography.

Watkins, Susan Cott, ed. *After Ellis Island: Newcomers and Natives in the 1910 Census.* New York: Russell Sage Foundation, 1994.

Notes

1. Readers can explore our findings at histmaps.research.cuny.edu. The mapping interface, called Map Explorer, gives users access to a wide variety of factors related to the growth of New York, including race and ethnicity, immigration, education, income, and population density.

2. Information on the National Historical Geographical Information System (NHGIS), which will be ongoing through at least 2006, can be found at www.nhgis.org. The New York study presented in this essay served as one of the precursors for the NHGIS. Beveridge is currently digitizing the other pre-1950 urban census data. Related analyses of New York include Andrew A. Beveridge and Susan Weber, "Shifting Patterns of Spatial Inequality: Race and Class in the Developing New York and Los Angeles Metropolises: 1940 to 2000," in David Halle, ed., *New York and Los Angeles: Politics, Society, and Culture* (Chicago: University of Chicago Press, in press, 2002), and Andrew A. Beveridge, "Immigrant Residence and Immigrant Neighborhoods in New York, 1910 and 1990," in Pyong Gap Min, ed., *Classical and Contemporary Mass Migration Periods: Similarities and Differences* (Walnut Creek, Calif.: Alta Mira Press, in press, 2002).

3. Joel Schwartz, "Tenements," in Kenneth T. Jackson, ed., *The Encyclopedia of New York City*, 1,161–63 (New Haven, Conn.: Yale University Press, 1995).

4. Walter Laidlaw, *Statistical Sources for Demographic Studies of Greater New York, 1920* (New York: The New York City 1920 Census Committee, Inc., 1922).

5. Another set of data applies to the period 1940 to 2000, based upon county-level data in the New York Metropolitan area. Unfortunately, tract-level data does not exist for the complete metropolitan area back to 1940. Indeed, due to census reporting changes, it is difficult to explore the changes from decade to decade. Such data does make it possible to assess the broad sweep of changes in the New York metropolitan area for seven censuses over sixty years.

6. Ilana Harlow, *The International Express* (New York: Queens Arts Council, 1995).

7. We assessed the degree of segregation with the dissimilarity index, the most common segregation index. It measures how evenly dispersed a group is with respect to the whole population or to other groups. It reports the proportion of a group that would have to be dispersed to make the group evenly distributed across all areal units—in this case, New York City census tracts. The higher the dissimilarity index, the more segregated the group. In 1910, the average level of segregation among the six largest origin groups was 0.54; in 1990 among the eleven largest groups in the metropolitan area, it was 0.75. In 1910, immigrants from Austria and Russia, many of whom were Jewish, had an index value of 0.39. Segregation levels were similarly low for Germans, English, and Irish. The Russians, however, were highly segregated from these long-standing groups, with index levels of 0.65 or higher. At the turn of the twenty-first century, immigrant segregation is more tied to race and language than to nationality or religion. The various West Indian groups are only moderately segregated from one another, while Germans are not highly segregated from Italians and Polish. Hispanic immigrant groups are highly segregated from one another and Asian groups even more so. For example, the segregation level between Dominicans and Colombians is 0.73, that between Chinese and Koreans 0.70. For more details, see Beveridge, "Immigrant Residence and Immigrant Neighborhoods."

CHAPTER SIX

REDLINING IN PHILADELPHIA

Amy Hillier

In the spring of 1933, with hundreds of homeowners facing foreclosures each day, President Franklin Roosevelt called on Congress to provide emergency assistance. "The broad interests of the Nation," Roosevelt said, "require that special safeguards should be thrown around home ownership as a guaranty of social and economic stability."[1] Congress responded by creating the Home Owners' Loan Corporation just two months later. The agency helped homeowners and private mortgage lenders alike by exchanging government bonds for defaulted mortgages on moderate-value homes. Between August 1933 and June 1936, the agency provided one million households across the country with new, low-interest, fifteen-year mortgages. Eight out of ten were able to save their homes and repay the agency in full.

In 1935, having made most of the loans it would make, the Home Owners' Loan Corporation and the Federal Home Loan Bank Board, its parent organization, developed the City Survey Program to investigate economic conditions, real estate trends, and racial and ethnic residential patterns in the nation's largest cities. Board members hoped that the survey would help them decide how to collect on the million outstanding loans, how to manage the sale of properties whose owners had defaulted on their new mortgages, and how to shore up the savings and loan industry by determining which lenders and communities needed federal support. Staff of the Home Owners' Loan Corporation, together with local realtors, lenders, and appraisers, generated written reports, detailed area descriptions, and color-coded "residential security" maps that indicated levels of risk to real estate investors for neighborhoods in 239 cities *(figure 1)*.

The board established national standards for grading residential areas and distributed an explanation of its standards to field staff across the country. In assigning areas a grade, field staff members were expected to consider demand for housing; homeownership rate; age and type of housing; social status of residents; adequacy of public utilities; access to schools, churches, businesses, and transportation; and presence of race-restrictive covenants aimed at maintaining homogeneity. First-grade areas, referred to as "A" or "best" and colored green, were expected to be racially and ethnically homogeneous and to have room for new residential growth. Second-grade or "B" areas, colored blue, were completely developed but were "still desirable." Third-grade or "C" areas, colored yellow, were "declining" and subject to "infiltration of a lower grade population." Fourth-grade or "D" areas were considered "hazardous" and colored red. They had lower homeownership rates, poor housing conditions, and an "undesirable population or an infiltration of it," referring largely to the presence of Jews and African Americans.

In the late 1970s, while conducting research for *Crabgrass Frontier*, urban historian Kenneth Jackson discovered the security maps in the archival records of the Federal Home Loan Bank Board. Jackson observed that red-colored areas in several cities corresponded with areas that had experienced massive disinvestment in the forty years since the maps were created. He argued that the Home Owners' Loan Corporation had caused redlining by sharing its maps with the Federal Housing Administration and private lenders who, in turn, avoided the red areas on the maps.[2] Researchers have traced the practice of mortgage discrimination back to at least the 1910s, but the word "redlining" was not used until the late 1960s, when community organizers in Chicago began identifying mortgage lenders and providers of homeowner insurance that drew red lines on maps around areas they refused to service.[3] There were few protections against redlining before the Fair Housing Act of 1968 outlawed discrimination at any

Figure 1. Residential security map for Philadelphia, 1937
Local real estate agents, appraisers, and lenders worked with staff from the Home Owners' Loan Corporation to create this security map for Philadelphia. The grades corresponded to their perception of real estate prospects, with red being reserved for hazardous areas. Records of the Federal Home Loan Bank Board, Record Group 95, Box 71, National Archives at College Park, College Park, Maryland.

stage in the home-buying process and the Home Mortgage Discrimination Act of 1975 mandated that certain financial institutions provide information about their lending practices. Jackson was the first to connect the maps created by the Home Owners' Loan Corporation with an old practice that was generating new attention at the time of his research.

In their subsequent efforts to explain the decline of central cities, urban researchers have elevated Jackson's redlining hypothesis to the status of fact while introducing little new evidence to support it. Writing about Gary, Indiana, Raymond Mohl and Neil Betten argued that the HOLC had a "pernicious impact" on segregation. They wrote that "The impact of the HOLC in Gary . . . was to consign the city's black sections, as well as adjacent white sections, to a future of physical decay and increased racial segregation." Lizabeth Cohen explained that in Chicago, "Faced with fewer alternatives after the depression to the big banks that respected these ratings, workers became victimized for years by a 'redlining' that originated with these HOLC classifications." Thomas Sugrue stated that the HOLC's residential security maps were the "primary sources used by brokers and lenders to determine eligibility for mortgages and home loans." People living in areas given "C" or "D" grades were "unlikely to qualify for mortgages and home loans. Builders and developers, likewise, could expect little or no financial backing if they chose to build in such risky neighborhoods."[4]

Writing about Charlotte, North Carolina, Thomas Hanchett argued that the HOLC survey caused disinvestment in low-income, mixed-use, and black areas and that it influenced decisions about lending in undeveloped areas. He focused on the agency's role in institutionalizing the already existing practice of redlining, a distinction that Jackson also made. Hanchett argued that "The HOLC's work served to solidify practices that had previously only existed informally," wiping out the "fuzziness" that existed when lenders determined creditworthiness on their own. "The handsomely printed map with its sharp-edged boundaries made the practice of deciding credit risk on the basis of neighborhood seem objective and put the weight of the U.S. government behind it."[5]

Rather than testing Jackson's theory that the Home Owners' Loan Corporation caused redlining, urban historians have endorsed and even embellished his account, extending his conclusions to a number of cities. I used GIS in conjunction with archival research to investigate the effect of the residential security maps on lending in

Philadelphia. Specifically, I used GIS to see where the Home Owners' Loan Corporation made its own mortgages, to understand how field agents assigned grades to Philadelphia's neighborhoods, and to determine if those grades affected private mortgage lending.

Home Owners' Loan Corporation records show that it made the majority of its loans in Philadelphia before the first security map of the city was drafted in 1935. So staff members could not have used the map grades to decide where to make loans. But analyzing the location of the agency's loans relative to the grades still allows for an assessment of its intent to provide assistance across racial and income groups. From agency loan summaries we know that in Chicago, Memphis, and Newark, 60 percent or more of its loans were made to areas it later gave third-grade (yellow) or fourth-grade (red) ratings. Since no summaries are available for Philadelphia, I collected and mapped data on individual loans there.

To map the location of loans using GIS, I digitized the final, 1937 version of the security map for Philadelphia and then *geocoded* a list of addresses where the agency made loans. Digitizing the security map was the more difficult of these tasks, although it was made easier by modifying a digital basemap rather than starting from scratch. The different colored areas on the security maps did not correspond to any political or administrative boundaries, so wards or census tract maps from the 1930s could not be used as basemaps. But the areas did correspond to streets, so I digitally combined census blocks to create the larger areas shown on the security maps *(figure 2)*.[6]

I then collected and geocoded the addresses for a random sample of three hundred loans that the agency made in Philadelphia. The security grade for the location of each mortgage was determined by joining the geocoded addresses and the digitized security map.[7] Results indicate that the attention the agency paid to areas it later deemed hazardous or declining was even greater in Philadelphia than in other cities.

Security grade	Loans	Percent of sample
First (green)	5	1.7%
Second (blue)	46	15.3%
Third (yellow)	63	21.0%
Fourth (red)	186	62.0%

Clearly, the agency did not refuse to make loans to homeowners in neighborhoods it deemed hazardous *(figure 3)*. Jackson and several other researchers have acknowledged this fact, but others have

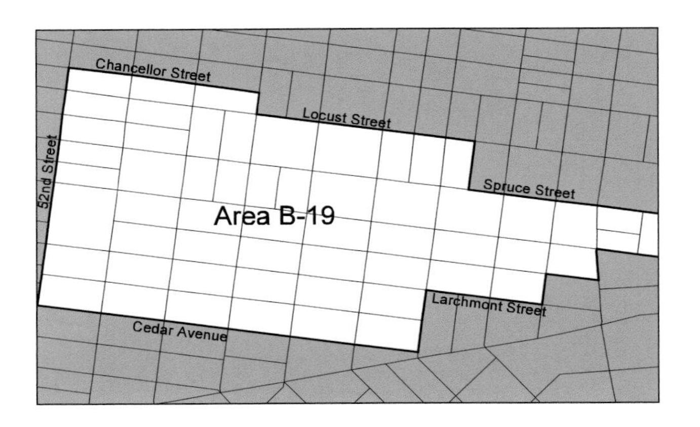

Figure 2
Once selected digitally, the boundaries of the blocks within this blue (second-grade or "B") area were dissolved to create ("digitize") Area B-19. Using an existing digital basemap made digitizing the residential security map simpler and more accurate.

failed to distinguish between the loans that the Home Owners' Loan Corporation made and the lending record of public and private lenders who might have had access to the security maps when they made decisions about loans. An argument that the Home Owners' Loan Corporation encouraged redlining must, therefore, focus on its later activities and the effect its maps had on subsequent lenders.

The Federal Home Loan Bank Board's materials were explicit about the basis for the security map grades, but the board left it up to the agency's staff and the real estate consultants to decide how to integrate and weight all of the different factors in order to assign each area a single grade. Of particular interest is the role of race in determining the map grade. Redlining generally refers to race-based lending discrimination, so learning whether race was a significant factor in determining the grade should tell us whether the Home Owners' Loan Corporation caused redlining.

Between 1935 and 1937, the Home Owners' Loan Corporation created three different drafts of the security map for Philadelphia *(figure 4)*. Changes from version to version show how staff refined their grading techniques. For example, later maps remove

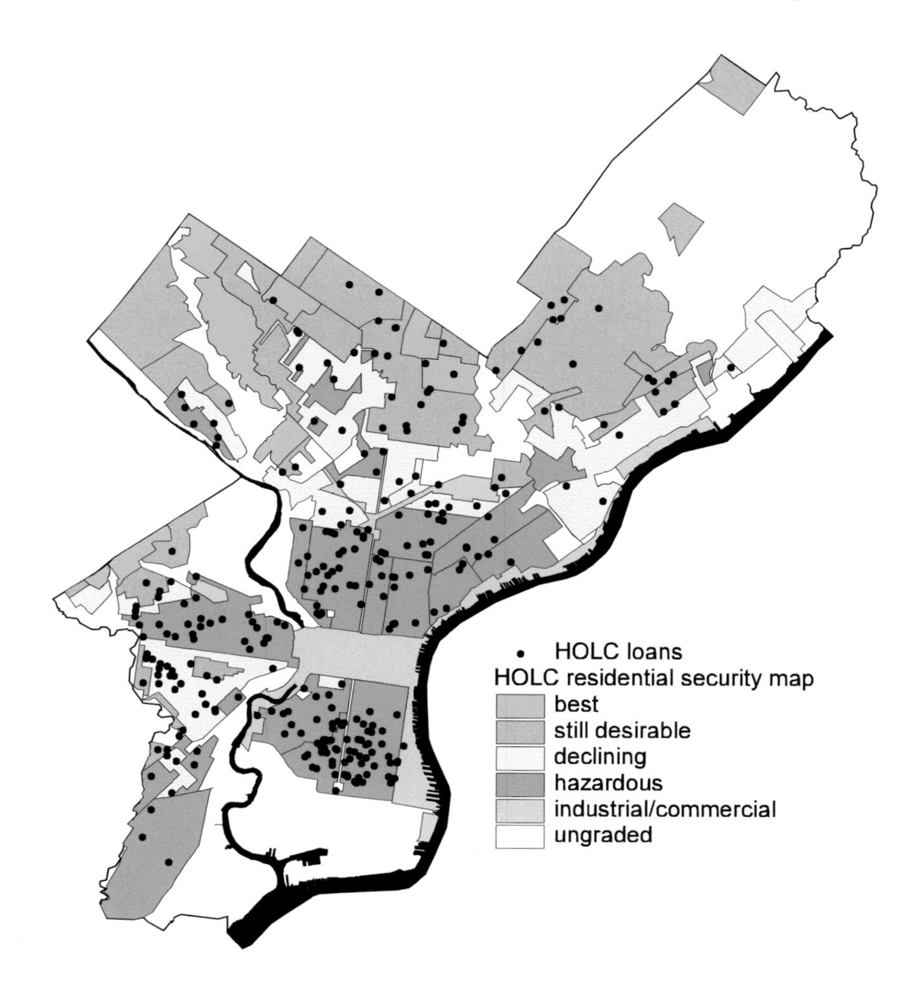

Figure 3. Home Owners' Loan Corporation loans in Philadelphia

Mapping the locations of a sample of loans made in Philadelphia shows that the agency provided a disproportionate amount of assistance to neighborhoods it deemed hazardous on the security maps.

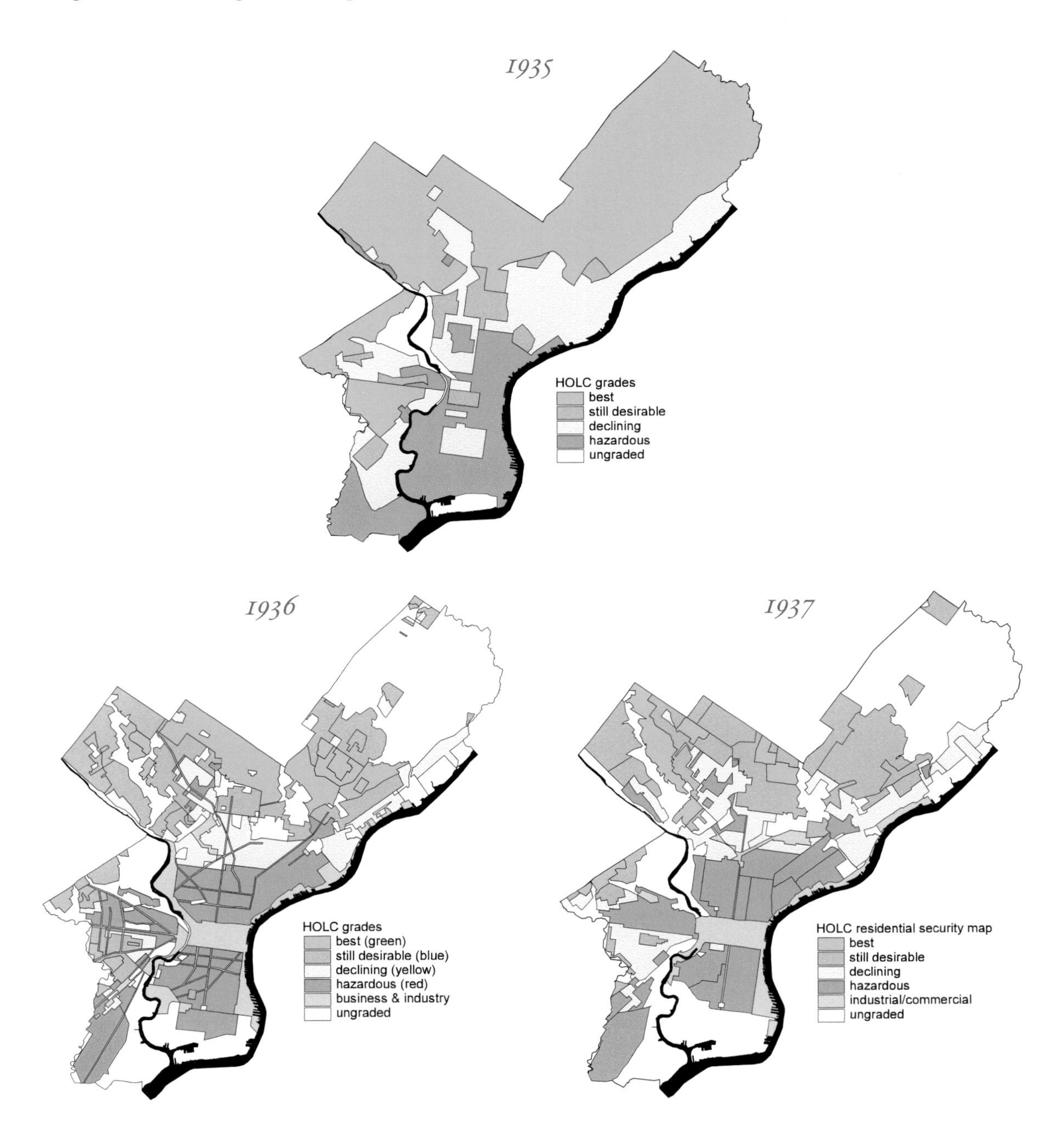

Figure 4. Successive drafts of the security map for Philadelphia, 1935–1937

The first draft map, drawn in 1935 (top), offered the most encouraging picture of real estate prospects in Philadelphia. Fifty-four percent of the total appraised area was coded green (best) while 18 percent was coded red (hazardous). In 1936 (lower left), conditions had changed little, but the agency assessed only 13 percent of real estate in the "best" category and red areas jumped to 31 percent. The gloom deepened in the 1937 map (lower right), which coded 34 percent of the city's real estate "hazardous" and only 8 percent "best."

grades from major parks, industrial and commercial corridors, and undeveloped land. They also demonstrate how the agency's perception of real estate prospects evolved in those three years. For although Philadelphia's housing and economic conditions changed very little, each successive map provided a gloomier picture of the city's real estate.

The City Survey Program reflected a larger shift in focus within the real estate and appraisal industries, from how creditworthy the property and borrowers were to how risky the neighborhood was. This change was partly intended to protect lenders against the kinds of losses they suffered during the Depression. It also reflected the influence of the ecological theory promoted by the Chicago School's Robert Park, Ernest Burgess, and Homer Hoyt, among others. This theory held that neighborhoods naturally decline as some residents move to find more suitable habitats. In this view, African Americans, Jews, and certain immigrant groups were seen as invaders, warning signs that neighborhoods had reached the last phase of their decline.[8] As long-term, self-amortizing mortgages became more popular, lenders were increasingly concerned about making loans in neighborhoods where decline seemed imminent or under way.

To understand the influence of ecological theory on Philadelphia's residential security map and, more specifically, to determine the effect of race, housing conditions, and location on the final security grade that areas received, I assigned each census tract a HOLC grade based on the proportionate area of the tract covered by each grade *(figure 5)*. It would be virtually impossible to calculate this as precisely without GIS software, which can automatically measure distance and area. Using tract-level data on property and residents from the 1934 Works Progress Administration's Real Property Survey and the 1940 U.S. Population Census, I confirmed that areas with more African Americans and more recent immigrants received poorer grades from HOLC staff, controlling for housing conditions, housing values,

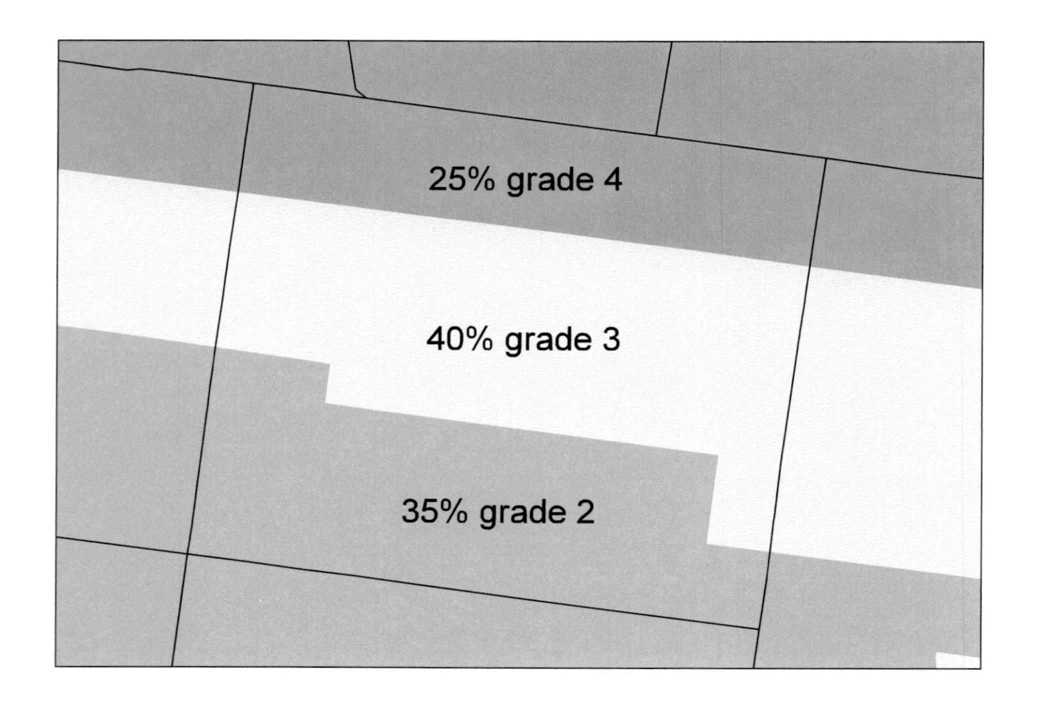

Figure 5. Assigning a grade to each tract

The areas defined on the residential security maps did not correspond to census tract boundaries. In order to assign a grade to each census tract, the proportion of the tract covered by each grade was calculated and multiplied by the grade. In this example, $(0.25 \times 4) + (0.40 \times 3) + (0.35 \times 2) = 2.9$, so this tract was assigned a grade of 2.9.

homeownership rate, and the location of the neighborhood within the city.[9]

The final question to ask about the Home Owners' Loan Corporation is whether the grades on the agency's security maps influenced where private lenders made their mortgages. This, rather than the agency's own lending record, was the basis for Jackson's charge that it caused redlining. Only if lenders avoided making loans on properties in areas marked red on the security maps or made loans with worse terms (such as higher interest rates or lower loan-to-value ratios) could the maps be said to have caused redlining. In order to test the redlining theory, then, I analyzed a sample of mortgage transactions in Philadelphia involving private lenders between 1937 and 1950 to see what effect the grades had on their lending patterns.

I collected information about a sample of loans (including a random citywide sample as well as all the loans made in four small areas) from the *Philadelphia Realty Directory and Service,* an annual listing of mortgage transactions published between 1926 and 1959 *(figures 6, 7).* I geocoded the sample properties, assigned them the appropriate security grade, and calculated the distance of each property from the boundary of a red area. Results show that neither the security grade nor the property's proximity to a red area explain differences in the total number of loans it received or in the loan-to-value ratio. Mortgages made in areas with worse grades and closer to red areas did have slightly higher interest rates. These higher interest rates probably reflected widespread knowledge of where racial minorities lived rather than access and adherence to the Home Owners' Loan Corporation grading scheme.

This empirical analysis indicates that urban researchers have overstated the significance of the security maps on lending, particularly by arguing that private lenders categorically refused to make loans in areas colored red. The areas considered most hazardous by the agency probably did suffer from disinvestment over the next

Figure 6. Philadelphia Realty Directory and Service
This directory listed all real estate transactions in Philadelphia, regardless of whether they involved a mortgage, organized by street address and the month and year of the transaction. For each transaction, the directory listed the owner, owner's address (or "P" if owner lived on premises), name of mortgage company (M), amount of mortgage, interest rate, and sale price (consideration). An index in the back of the directory included a complete list of properties, their size, assessed value, and date of the most recent transaction.

7

PROPERTY	GRANTEE OR MORTGAGEE	ADDRESS	MTGE	CONS OR INT RATE
		OAKMONT ST (Contd)		
4313	Chester Smith etux	Premises		4,490
"	(M)Phila Sav Fund Soc		4,000	4½%
4315	Wm Banaszak etux	Premises		4,490
"	(M)Phila Sav Fund Soc		4,000	4½%
4317	Joseph A Morris etux	Premises		4,490
"	(M)Phila Sav Fund Soc		4,000	4½%
4319	Clifford J Adair etux	Premises		4,490
"	(M)Phila Sav Fund Soc		4,000	4½%
4327	Wm Furey etux	Premises		4,490
"	(M)Phila Sav Fund Soc		4,000	4½%
4343	Bernard H Forsting etux	Premises		4,490
"	(M)Phila Sav Fund Soc		4,000	4½%
4747	Allen R Coffin etux	506 E Roumford St		2,800
		OGDEN ST		
Ss 45'W of Carlisle St	James F Hickey	200 E Price St		224,054(B)
"	James V Ross	1430 Spruce St		Nom
"	James F Hickey	200 E Price St		Nom
1424	James F Hickey	200 E Price St		224,054(B)
"	James V Ross	1430 Spruce St		Nom
"	James F Hickey	200 E Price St		Nom
1420-22	James F Hickey	200 E Price St		807,750(B)
4929	Chas R Frederick etux	Premises		1,350
5106	Jos Miller etal	758 S 4th St		Nom
5128	Thomas F Cullen etux	Premises		1,750
"	(M)West Phila Fed S&L		1,350	6%
5145	Rose Paul	23rd&McKean Sts		500(B)
5321	Lillian M Zerbey (Howard)	Premises		1,300

several decades, as Jackson noted. But disinvestment happened independently of the HOLC's security maps. Rather than causing redlining, the maps reflected the conditions in Philadelphia's neighborhoods in the 1930s as well as the dominant attitudes and methods of the real estate and appraisal industries.

Redlining during the middle decades of the twentieth century was a more complicated process than many historians have appreciated, in part because the Home Owners' Loan Corporation was neither the only nor the first lending organization to make maps with symbolic red lines. The Federal Housing Administration (FHA), created in 1934 to protect mortgage lenders against the risk of foreclosure, started collecting quantitative data and making maps a year before the Home Owners' Loan Corporation initiated its City Survey program *(figure 8)*.

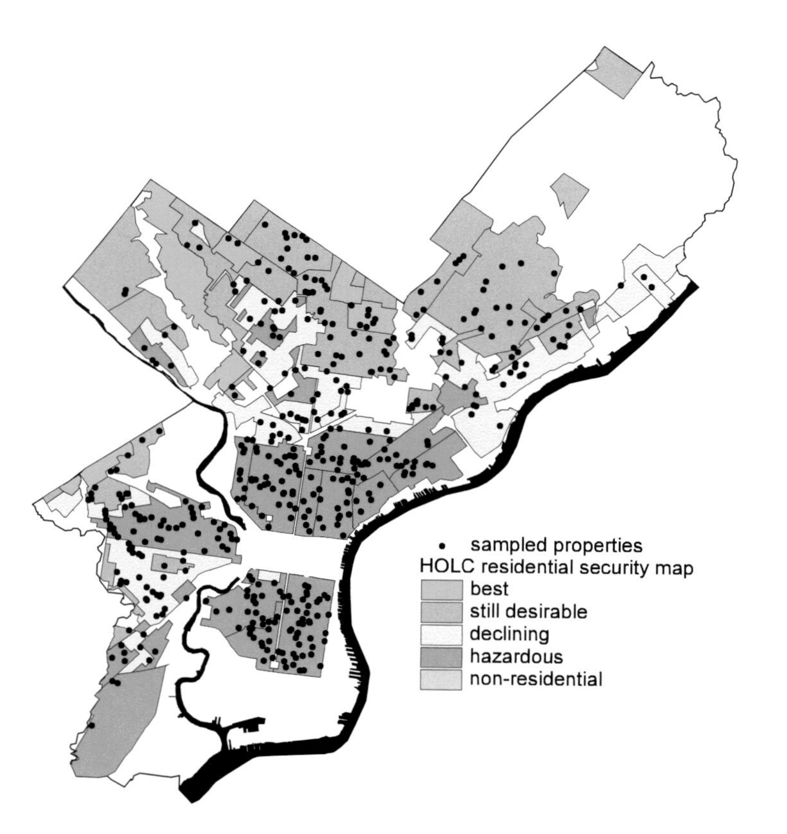

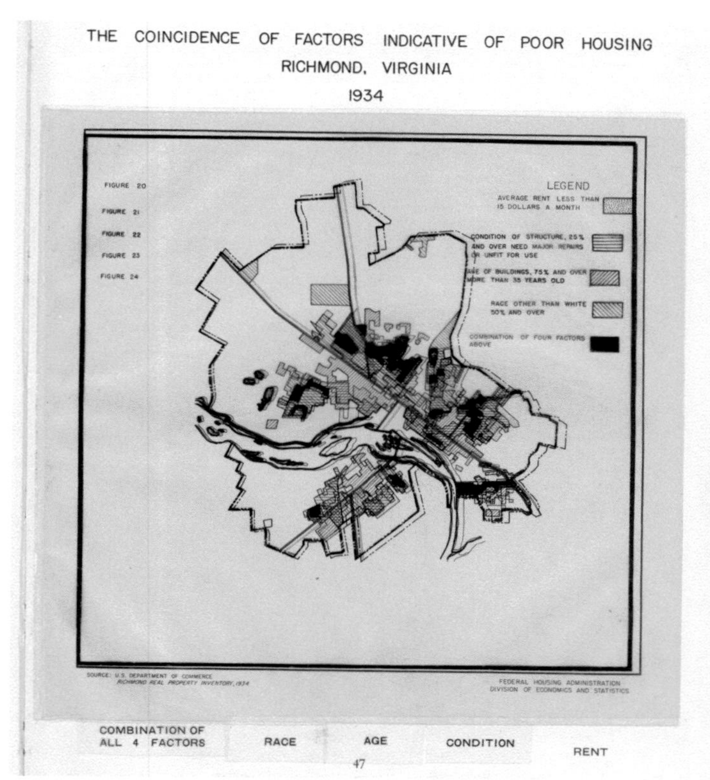

Figure 7. Random citywide sample of mortgages made by private lenders in Philadelphia, 1937–1950

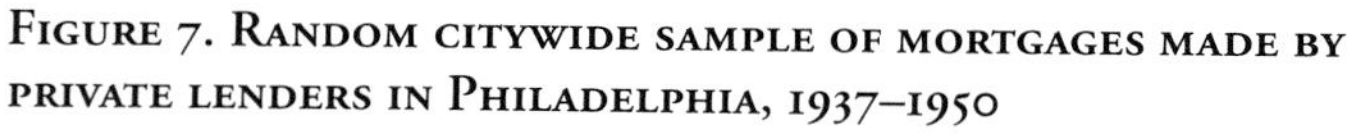

Using the Philadelphia Realty Directory and Service, a random sample of properties was selected. Properties in the central part of the city, most of which was colored red, made up the majority of this sample. Data was collected on all of the transactions at each of these five hundred properties and analyzed in conjunction with the 1937 version of the HOLC's map.

Figure 8. FHA overlay map

The FHA included this map in The Structure and Growth of Residential Neighborhoods in American Cities in 1939. The map used a series of transparent overlays to represent map layers in order to determine the spatial concentration of undesirable housing and demographic characteristics.

The FHA created and shared detailed maps showing racial composition and housing conditions in cities across the country. It also promoted its own standards for appraising neighborhood risk levels by requiring that private lenders follow them in order to receive federal mortgage insurance.

The FHA, like the Federal Home Loan Bank Board, also encouraged private lenders to make their own maps, although private lenders probably did not need any encouragement. The former chief appraiser for the Metropolitan Life Insurance Company in Philadelphia, J. M. Brewer, created a map of the city that categorized neighborhoods according to class and showed where Jews, Italians, and "Colored" people lived *(figure 9)*. He did this a year before the Home Owners' Loan Corporation created its first security map for Philadelphia—and

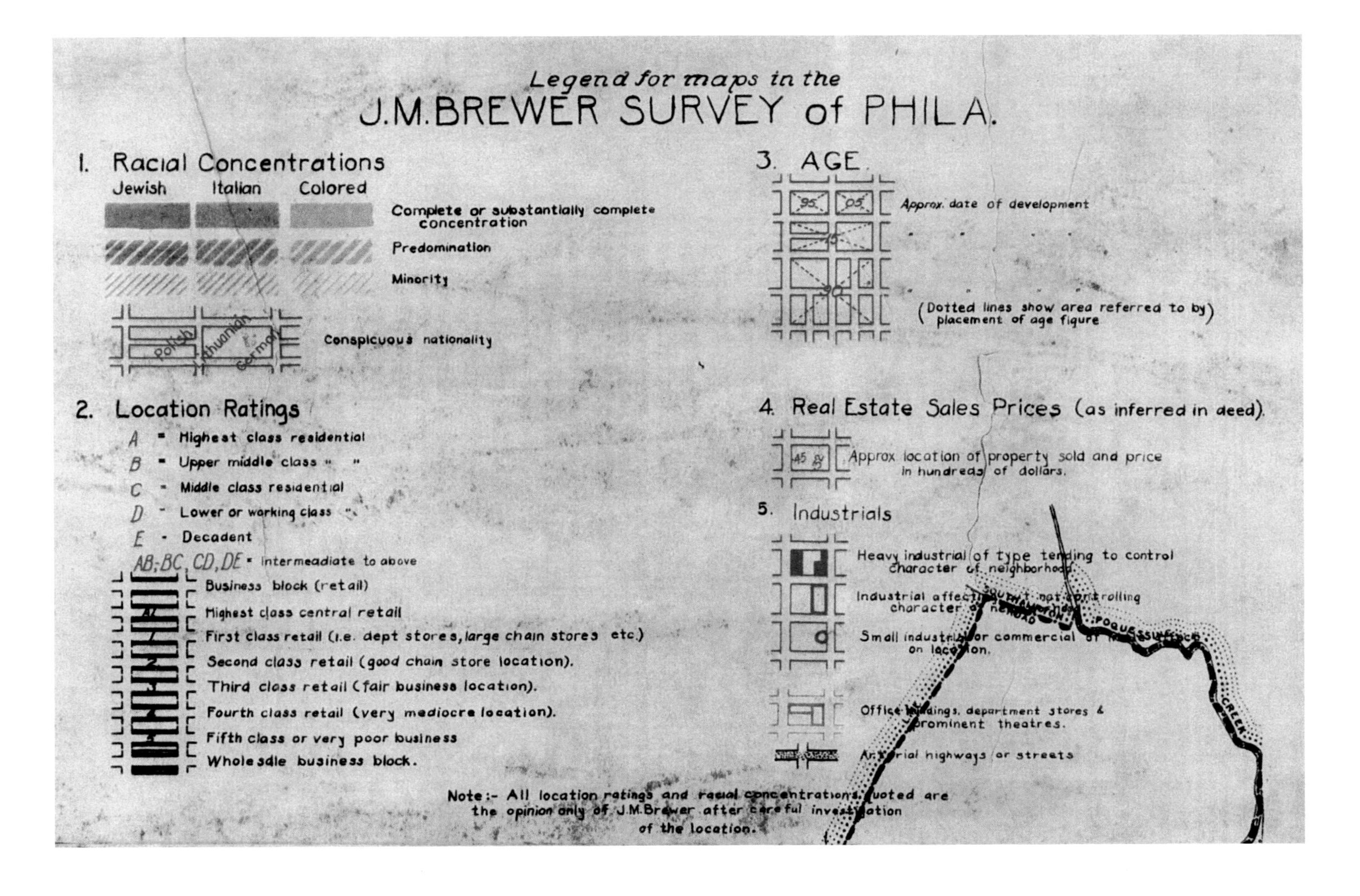

Figure 9. Legend of J. M. Brewer's map of Philadelphia
Brewer conducted his detailed block-level survey in 1934, before the start of the City Survey Program. Brewer later served as map consultant to the Home Owners' Loan Corporation. Map Collection, Free Library of Philadelphia, Philadelphia, Pennsylvania.

he later served as a map consultant to the agency. The Security-First National Bank of Los Angeles also had an active research staff that produced a map of Los Angeles neighborhoods displaying five residential categories that borrowed the language of ecological theory: subdivision, growth, maturity, decline, or decadence *(figure 10)*. Although the bank created this map after the Home Owners' Loan Corporation started its maps in Los Angeles, the Security-First classification system was different enough to indicate that it was made separately. While the Philadelphia and Los Angeles maps created by private organizations used different colors and categories, they assumed a racial and ecological conception of neighborhood change just as the Home Owners' Loan Corporation's security maps and FHA's underwriting guidelines did.

GIS helps one see the spatial patterns that constituted redlining. It brings a precision to the analysis that many previous studies have lacked. As researchers discover more maps that lenders created and examine mortgage lending patterns in other cities, GIS offers them a valuable complement to traditional archival research for understanding where lenders made loans and how they decided where to lend.

Figure 10. Security-First National Bank of Los Angeles real estate map
The FHA printed this black-and-white version in its journal, Insured Mortgage Portfolio, in 1938.

Acknowledgments

The work that provided the basis for this research was supported by funding under a dissertation grant from the U.S. Department of Housing and Urban Development. The substance and findings of that work are dedicated to the public. The author is solely responsible for the accuracy of the statements and interpretations and such interpretations do not necessarily reflect the views of the U.S. Government. This research was also supported by a grant from the Research Institute for Housing America. I would like to thank Dennis Culhane, Georgette Poindexter, Thomas Sugrue, and Tony E. Smith for their assistance with the dissertation upon which this chapter is based.

Hardware and software

Pentium® II PC with 64 megabytes RAM; ArcView 3.1.

Further reading

Bailey, Trevor C., and Anthony C. Gatrell. *Interactive Spatial Data Analysis.* Essex, England: Longman Scientific & Technical, 1995.

Bartelt, David. "Housing the 'Underclass.'" In Michael Katz, ed. *The "Underclass" Debate: Views from History.* Princeton: Princeton University Press, 1993.

Fix, Michael, and Raymond Struyk, eds. *Clear and Convincing Evidence: Measurement of Discrimination in America.* Washington, D.C.: The Urban Institute Press, 1993.

Goering, John, and Ron Wienk, eds. *Mortgage Lending, Racial Discrimination, and Federal Policy.* Washington, D.C.: The Urban Institute Press, 1996.

Harriss, C. Lowell. *History and Policies of the Home Owners' Loan Corporation.* New York: National Bureau of Economic Research, Inc., 1951.

Hoagland, Henry E., and Leo D. Stone. *Real Estate Finance.* Homewood, Ill.: Richard D. Irwin, Inc, 1961.

Hoyt, Homer. *The Structure and Growth of Residential Neighborhoods in American Cities.* Washington, D.C.: Federal Housing Administration, 1939.

Massey, Douglas S., and Nancy A. Denton. *American Apartheid.* Cambridge, Mass.: Harvard University Press, 1993.

Metzger, John T. "Planned Abandonment: The Neighborhood Life-Cycle Theory and National Urban Policy." *Housing Policy Debate* 11, no. 1 (2000): 7–40.

Mohl, Raymond A. "Trouble in Paradise: Race and Housing in Miami during the New Deal Era." *Prologue* 19 (1987): 7–21.

Squires, Gregory, ed. *Insurance Redlining: Disinvestment, Reinvestment, and the Evolving Role of Financial Institutions.* Washington, D.C.: The Urban Institute Press, 1997.

Notes

1. David A. Bridewell, *The Federal Home Loan Bank Board and Its Agencies* (manuscript in the Research Library of the Federal Home Loan Bank Board, dated May 14, 1938), 186.
2. Kenneth Jackson, *Crabgrass Frontier* (New York: Oxford University Press, 1985), 190–218.
3. John Goering and Ron Wienk, eds., *Mortgage Lending, Racial Discrimination, and Federal Policy* (Washington, D.C.: The Urban Institute Press, 1996); Gregory Squires, *From Redlining to Reinvestment: Community Responses to Urban Disinvestment* (Philadelphia: Temple University Press, 1992).
4. Raymond A. Mohl and Neil Betten, *Steel City: Urban and Ethnic Patterns in Gary, Indiana, 1906–1950* (New York: Holmes & Meier, 1986), 66–70; Lizabeth Cohen, *Making a New Deal: Industrial Workers in Chicago, 1919–1939* (New York: Cambridge University Press, 1990), 276; Thomas J. Sugrue, *The Origins of the Urban Crisis* (Princeton, N.J.: Princeton University Press, 1996), 43–44.
5. Thomas W. Hanchett, *Sorting Out the New South City: Race, Class, and Urban Development in Charlotte, 1875–1975* (Chapel Hill: University of North Carolina Press, 1998), 231.
6. Census blocks represent the space created by intersecting streets. Census tracts are made up of block groups, which are made up of census blocks.
7. Geocoding assigns x and y coordinates to addresses located along line segments (each representing a range of house numbers) within a street centerline file. Through a spatial join, geocoded addresses can be assigned the attributes of a geographic area, such as areas defined by the HOLC, that they fall into.
8. Robert E. Park and Ernest W. Burgess, *The City: Suggestions for Investigation of Human Behavior in the Urban Environment* (Chicago: The University of Chicago Press, 1925); Homer Hoyt, *The Structure and Growth of Residential Neighborhoods in American Cities* (Washington, D.C.: Federal Housing Administration, 1939); Jackson, "Race, Ethnicity, and Real Estate Appraisal: The Home Owners' Loan Corporation and the Federal Housing Administration," *Journal of Urban History* 6 (August 1980): 423–24; Raymond Moyl, "The Second Ghetto," in June Manning Thomas and Marsha Ritzdorf, eds., *Urban Planning and the African American Community: In the Shadows* (Thousand Oaks, Calif.: Sage Publications, 1997).
9. All relationships were significant at the 0.05 level. This statistical analysis used a spatial lag model, rather than Ordinary Least Squares regression, in order to incorporate spatial autocorrelation, the spatial dependence within values on the dependent variable that can lead to biased estimates of the size and significance of relationships. The weight matrix used defined neighbors as all census tracts with centroids within 0.6 miles. The model explained an estimated 59 percent of the variance in HOLC grade. See Amy Hillier, "Redlining and the Home Owners' Loan Corporation" (*Journal of Urban History,* forthcoming).

CHAPTER SEVEN

CAUSES OF THE DUST BOWL

Geoff Cunfer

When the southern plains erupted in dust storms in the middle 1930s, observers immediately began searching for explanations. For a decade, agricultural economists had argued that cropping submarginal land in the Great Plains was the cause of widespread rural poverty. Now they believed that cropping land fit only for grazing was causing dust storms, too. An effort to classify land according to its proper use and to "adjust" land use on submarginal tracts was already underway when Franklin Roosevelt took office in the depths of the Great Depression. Roosevelt incorporated this effort into his New Deal government and extended to it broad license to adjust farming. Just as that work began in Washington, drought and dust storms descended on the Great Plains *(figure 1)*. Newly formed federal initiatives like the National Resources Board, the Land Utilization Program, the Resettlement Administration, the Soil Conservation Service, and the Farm Security Administration argued that not only did misuse of land result in abject poverty for millions of plains residents, it was now causing disastrous dust storms that carried across half a

Figure 1
A dust storm descends on Baca County, Colorado, c. 1936. Photograph by D. L. Kernodle. Library of Congress, Prints and Photographs Division, FSA-OWI Collection, LC-USZ62-013580.

continent. New Deal officials told a story of decline and environmental mismanagement: homesteaders who flooded the region around the turn of the century had plowed land fit only for grazing, and in so doing had created the Dust Bowl, with all of its social, economic, and environmental disruption.[1]

People living in the plains hated this story. Chambers of Commerce, town boosters, and newspaper editors protested the suggestion that they were farming unfit land, that they should stop doing so, and that the region could not support current populations.[2] Kansas historian James Malin argued vigorously that dust storms were a routine part of plains life, that the drought would pass, and that the region could continue its current way of farming. But the New Deal story was powerful, strengthened by a brilliant public relations campaign that sent skilled photographers like Dorothea Lange and Arthur Rothstein to the plains to document its misery *(figure 2)* and generated evocative films like Pare Lorenz's "The Plow that Broke the Plains." By the 1940s, rain had returned to the plains, the economy had recovered, and Americans turned their attention to war. Time moved on and times were better. Yet the New Deal story of the 1930s remained fixed in the popular consciousness: the dust storms had been caused by misuse of land, and that is the explanation still taught by most environmental historians.

Around 1980 three books evaluating the Dust Bowl appeared. The most influential was *Dust Bowl: The Southern Plains in the 1930s,* by Donald Worster, an eloquent history that reinvigorated the New Deal story, and which has become required reading in many university history classes. According to Worster, capitalism drove farmers to plow up too much land, causing the Dust Bowl.[3] R. Douglas Hurt, in *The Dust Bowl: An Agricultural and Social History,* identifies several causes of dust storms—erodible soils, drought, high winds, and technological innovations that allowed farmers to plow more land than before. But his main focus is on the social and economic pain suffered by people in the region. Mathew Bonnifield, in *The Dust Bowl: Men, Dirt, and Depression,* outlines the government's response to the

Figure 2
Heavy black clouds of dust rising over the Texas Panhandle, March, 1936. Photograph by Arthur Rothstein. Library of Congress, Prints and Photographs Division, FSA-OWI Collection, LC-USZ62-125986.

crisis without clearly addressing the cause of the dust storms.

This chapter asks the same question posed by New Dealers during the Depression and by Worster and Hurt: what natural and human factors contributed to the dust storms of the 1930s? It uses different methods and evidence than earlier studies and explores the question at a regional scale. Worster explains the Dust Bowl through intensive case studies of two counties, relying on New Deal government documents, newspaper articles, and interviews with residents. In contrast, this study evaluates all 280 counties in the Dust Bowl region, beginning with the scientific literature about the causes of wind erosion, then analyzing data from agricultural censuses, soil surveys, and weather stations using geographic information systems (GIS) choropleth and overlay mapping. *Choropleth* mapping portrays quantities for areal units (in this case counties) by varying color, while *overlay* mapping superimposes one set of map features over another to reveal spatial relationships. Analyzing information about all the Dust Bowl counties has the advantage of being comprehensive while allowing a fairly high level of detail. It does not, however, account for individual farms or communities, thus sacrificing some precision for systematic and comprehensive coverage.

GIS allows environmental historians to test systematically the extent to which detailed case studies apply to broader regions. Here GIS links data about soil type, land use, and weather to each county to portray geographical patterns. This study uses five map layers. One represents county boundaries in 1930. The other four chart the locations of dust storms at different times during the 1930s. The Dust Bowl moved around from year to year, and figure 3 traces those shifting locations.[4] The first dust region outline (brown) indicates the location of the worst wind erosion in 1935 and 1936. The second (green) shows dust storm locations in 1938, and a third (blue) presents the 1940 dust region. A fourth outline (red) represents the core of the Dust Bowl between 1935 and 1938. During the decade, the Dust Bowl moved

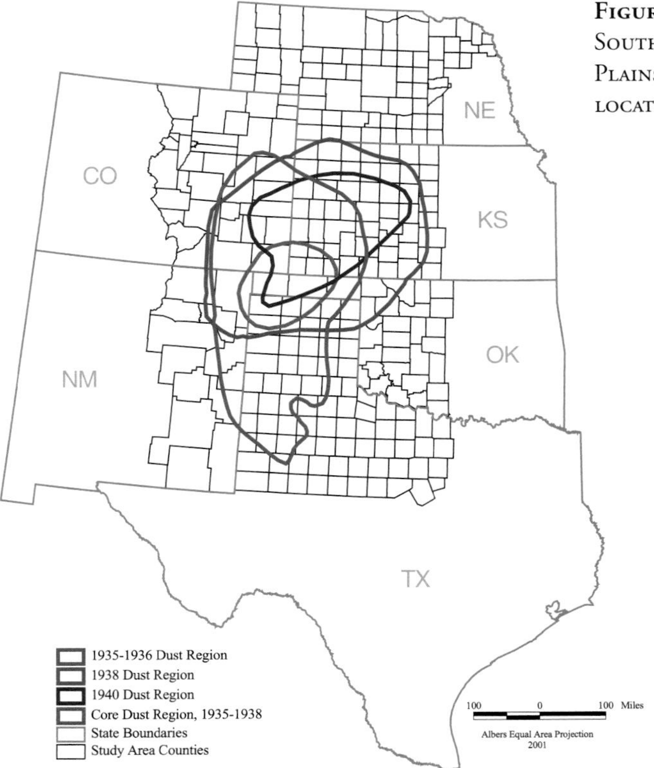

Figure 3
Southern and central Great Plains counties and dust storm locations, 1935–1940.

north and contracted, from brown to green to blue in figure 3.

To understand why dust storms happened where and when they did it is essential to understand the physics of wind erosion.[5] Erosion begins when strong winds blow across dry soil. As wind increases beyond "threshold velocity," soil begins to move. Medium-sized soil particles move first, then eventually dislodge both smaller and larger particles. Soil bounces along the soil surface in long, flat arcs, in a process called "saltation." Saltation, akin to sandblasting, sets everything else in motion. Saltating pieces strike larger objects that would normally resist the wind, pushing them, sliding or rolling, along the surface. Mid-sized particles also pulverize the larger pieces they hit, breaking them into medium-sized particles that then begin saltating in their own right. They also strike smaller particles, kicking them up into the air in suspension. Fine, suspended dust is the only type that moves long distances in dust storms.

Large, sturdy objects, of course, resist the wind—well-rooted plants, stones, standing grain stubble, and large clods of dirt held together by moisture. The soils most resistant to blowing are the loams, because they tend to form clods. The erodibility of heavy clays varies from season to season, while sandy soils erode most readily. Since erosion varies as the cube of wind velocity and inversely as the square of soil moisture, a little wind beyond threshold velocity can move a lot more soil, while a little moisture can prevent much erosion. Altering soil moisture is the most effective means of changing threshold velocity. A dry field may take winds of only twelve miles per hour to begin blowing, while after a light rain the threshold velocity in the same field rises to thirty miles per hour. The amount of moisture in any field is the result of many factors, but rain is most important, followed by temperature, since hot days quickly dry soils. Sound pastures and fields of well-established crops thriving in midsummer are generally safe from wind erosion. Most cropped fields are bare of living vegetation, however, during the times of high erosion risk in late winter and spring.

Dust storms are most common in the spring, infrequent in fall, and rare in summer and winter. March sees the most dust storms, followed by April and February. In spring, winds are at their strongest and the southern landscape is thawed and free of snow, but little new vegetation has yet emerged. In winter, the ground is frozen and winds are mild. By summer, plant growth protects the soil. But in between, the right conditions can lead to dust storms.

The Great Plains Population and Environment Project has interpolated soil data collected in the State Soil Geographic Data

Base (STATSGO) to the county level (see Data sources). Figure 4 shows the percent of the soil cover in each county that is sandy. The darkest counties have 60 percent sandy soils or more. They are the most likely to erode on the basis of soil texture alone. The red outline shows the area of the worst dust storms during the four-year period 1935–1938. All of the counties within the outline are moderately sandy, but the sand hills of north-central Nebraska and the southern end of the Texas–New Mexico border, where sandiness tops 60 percent, were not prominent in the Dust Bowl. Figure 4 suggests that while the sandiness of soils contributed to dust storms, it was not overwhelmingly important.

The more land that farmers plow in a county, the more exposed soil there will be in March, the peak of the dust season. Figure 5 shows the percentage of land plowed for crops in each county in the decade leading up to the dust storms of 1935–1936. Counties in white had 10 percent of their area cropped or less, leaving 90 percent or more of their land in grazed but otherwise undisturbed native grass. The darkest brown counties were heavily farmed, with 60 to 82 percent of their land in crops. An expansion of farming between 1925 and 1935 is evident in the maps. While the counties on the eastern edge of the plains were already as plowed as they would get by 1925, farmers in western plains counties continued to break new sod into the early 1930s, and many counties there moved up one or two categories during the decade before 1935.

The black outline in figure 5, marking the 1935–1936 dust region, shows little relationship between dust storms and counties with large amounts of cropland. Some counties subject to dust storms did indeed have increased acreage in crops just prior to 1935. They included the two counties Donald Worster highlighted: Cimarron County, Oklahoma, which tripled its crop acreage between 1925 and 1935, and Haskell County, Kansas, which doubled its land in crops. But areas in central Kansas and Nebraska with much cropland experienced few dust storms in those years. Dust storms did occur in counties that had very little crop land, including eleven counties with at least 80 percent native

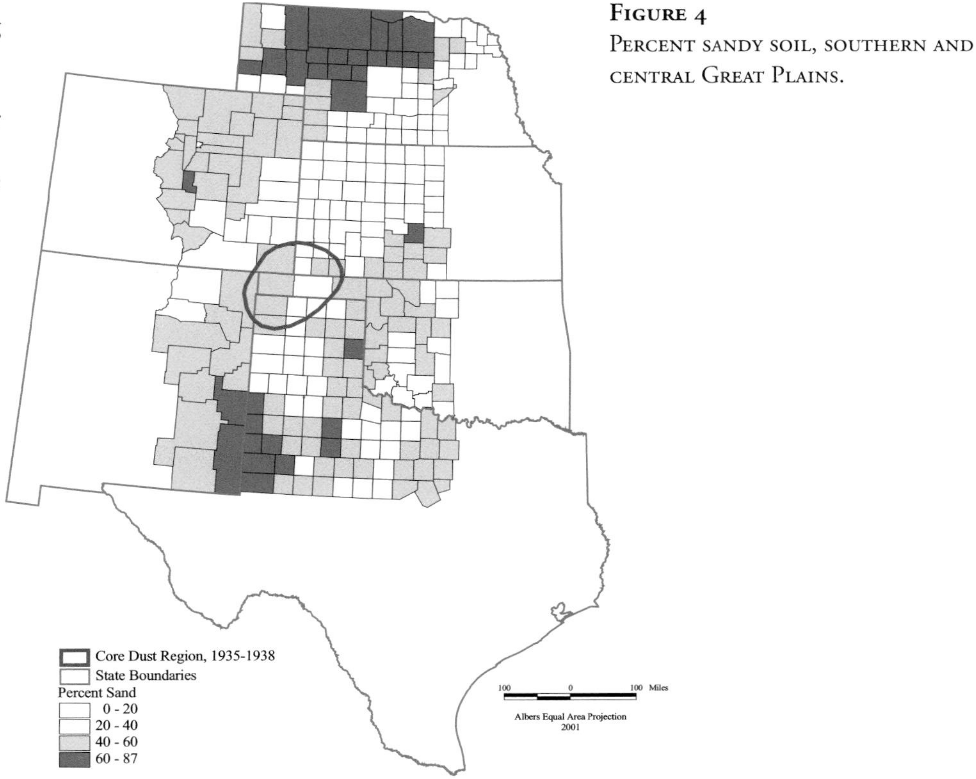

Figure 4
Percent sandy soil, southern and central Great Plains.

grassland and another three with at least 90 percent grass cover. Those fourteen ranching counties were in the western part of the Dust Bowl, upwind of the more plowed-out eastern and southern plains. New Deal activists and Worster laid blame for the Dust Bowl firmly on misuse of the land by grain farmers. This set of maps is much more ambiguous.

Drought explains the location of dust storms better than land use. Figure 6 maps rainfall in the southern and central plains in a sequence of maps that show the drought and the dust storms moving roughly in tandem from year to year. Each item shows a dust region overlaid on a map of county rain shortage for the five years prior to that dust season. Only a handful of counties (mapped in blue) had more rain than average between 1932 and 1940. And few counties fall into the 0 to –10 percent category. Most counties were 10 percent drier than average, or more, and more than eighty counties were at least 20 percent drier. The drought was deep, extensive, and persisted for nearly a decade. It was the driest time farmers on the plains have faced since the 1890s, when systematic weather monitoring began there. The severe drought of the era might better explain the occurrence of

Figure 5
Percent of total county area devoted to crops compared to 1935–1936 dust storm region.

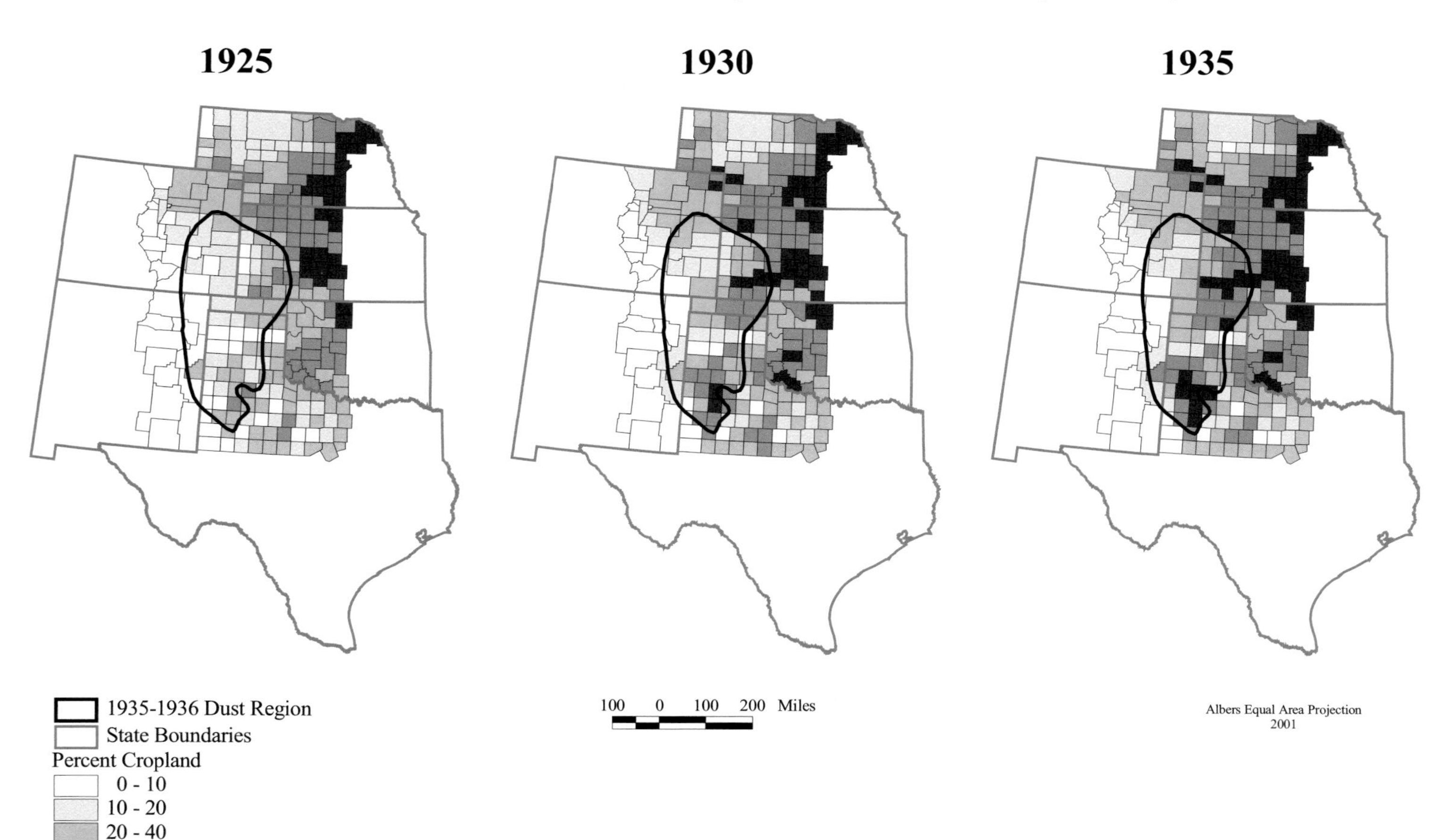

dust storms than the land-use practices of plains farmers.

Other research suggests that high temperatures increase the incidence of dust storms.[6] Since the 1960s (when good data on blowing dust becomes available) dust storms have been a problem only on the southern and central plains, where higher temperatures increased evaporation. Southern warmth also meant that there was little snow cover and unfrozen soils during the windy months of February and March. Dust storms happen most frequently on the southern plains, occasionally on the central plains, and rarely on the northern plains. Temperature best explains this, since rainfall, land use, and soils are roughly similar north to south. Figure 7 maps the difference of average March high temperatures from normal. The plains were dry in the 1930s, and they were hot, too. In 1935 and 1936 every county in the region was at least 3 percent warmer than usual in March, and nearly half of the counties were more than 10 percent warmer. In 1938 only a handful of counties were cooler than normal, while in 1940 only Nebraska and parts of eastern Colorado were slightly cooler than usual. March warmth melted snow early, thawed soils early, and dried the ground, contributing

Figure 6
Percent difference from average rainfall for five-year periods preceding dust seasons.

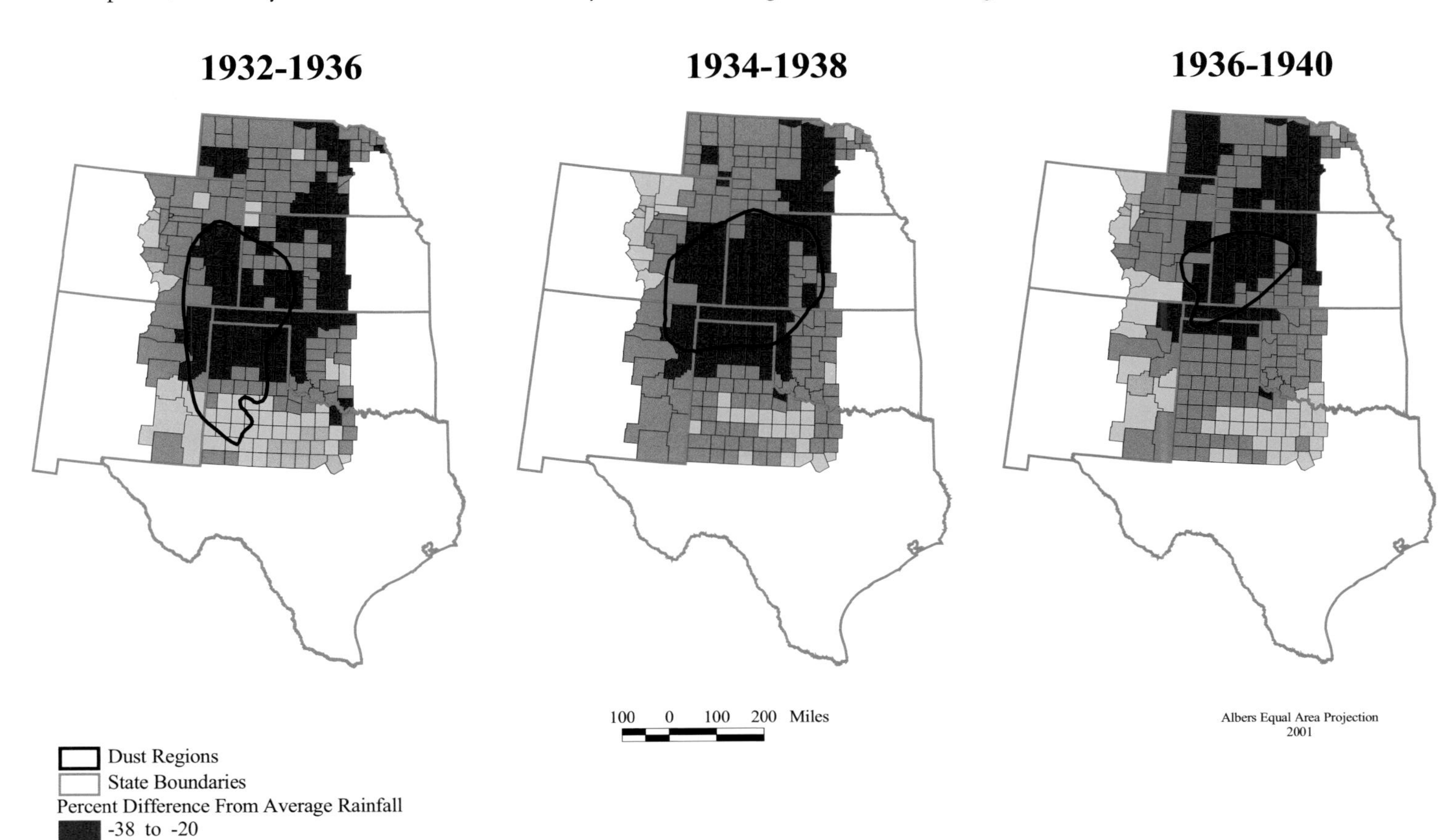

to the increase in dust storms and to their extension farther north in the 1930s than in other eras. But the March heat waves do not match dust storm regions as closely as drought. Dust storms certainly happened in places that were noticeably warmer than average, but not in all of them.

Several important causes of dust storms do not appear in this study. Monthly wind data since the 1890s is not yet available for county-level analysis, but soon will be.[7] Soil surface texture, which affects erodibility, can only be measured effectively at the scale of the individual field. The length of an exposed field in the direction of the wind is also important because erosion grows exponentially with distance, but field lengths cannot be adequately evaluated at the county level. Finally, overgrazing by livestock on unplowed rangeland may have contributed to soil exposure and thus to dust storms. This attempt to understand the causes of dust storms at a regional scale cannot be complete. It does, however, reveal important connections between many known causes of wind erosion and the location of dust storms in the 1930s.

The maps presented here suggest a prominent role for drought in causing the Dust Bowl. Plowing for crops certainly exposed

Figure 7
Percent difference from average March high temperature compared to dust storm regions.

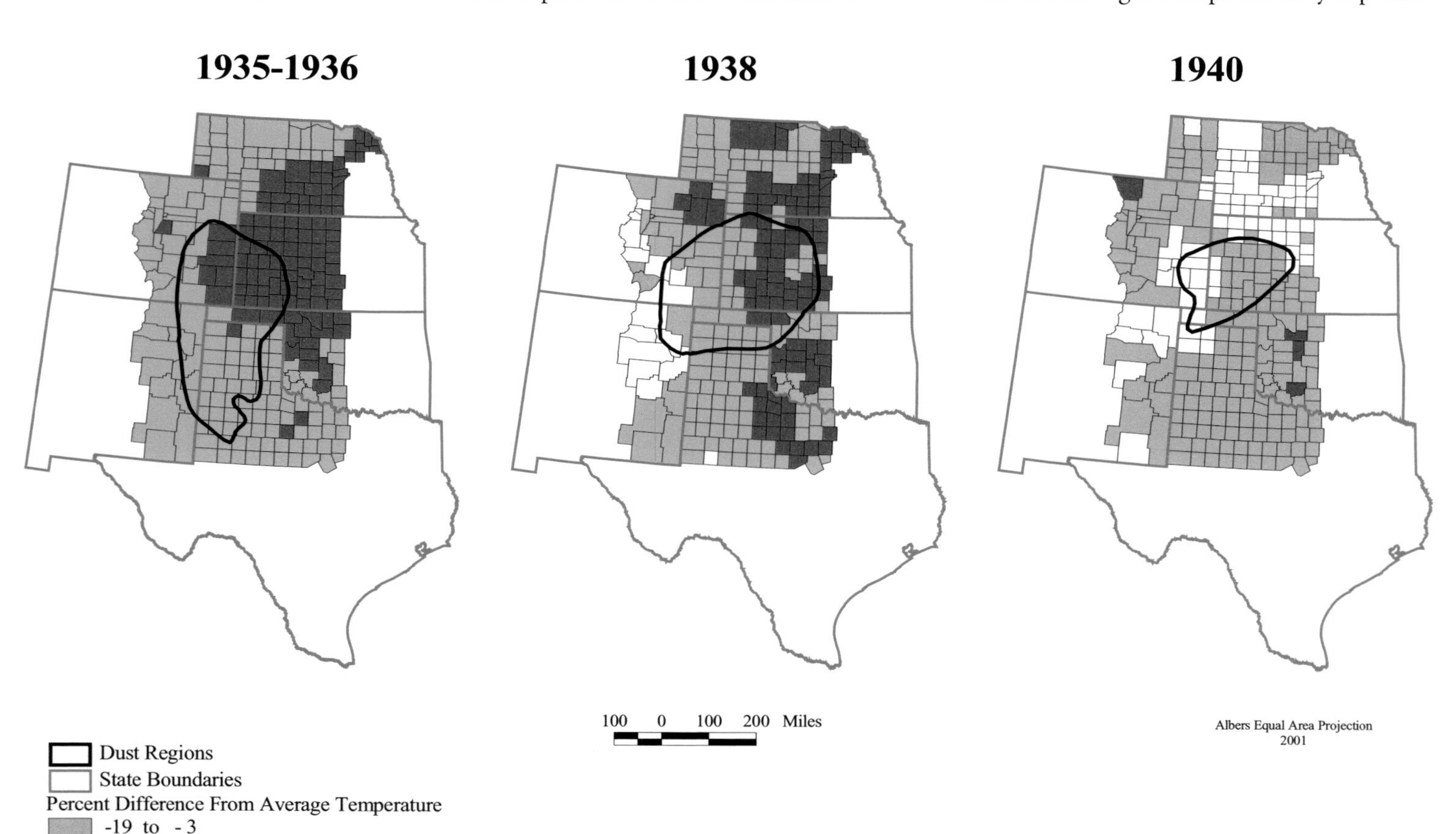

land to erosion, and may have tipped the balance in the southern Texas panhandle in 1935 and 1936, but this study suggests that the ways people used the land had less to do with creating dust storms than did the weather. Figure 5 shows that tens of millions of acres of land in the plains were not plowed for crops, and that some of that land experienced dust storms nonetheless. These results agree with some earlier studies. An examination of dust activity at plains weather stations from 1961 through 1988 found a similarly prominent role for low rainfall and also for high temperatures, while land-use practices had less explanatory power.[8] James Malin documented hundreds of dust storms in Kansas between 1850 and 1900, long before farmers had plowed more than a small fraction of the prairie sod *(figure 8)*. Southern plains archaeologists have found many episodes of Native American occupation, each separated from others by deep layers of wind-deposited soil.[9] It is time for environmental historians to consider the possibility that dust storms are a normal ecological disturbance that coincides with extended periods of drought and high temperatures on the southern plains, rather than evidence of human ecological failure.

Figure 8
A dust storm passes over Midland, Texas, February 20, 1894, when more than 96 percent of the Texas panhandle remained in unplowed, native grass. Photograph by H. G. Symonds. National Archives and Records Administration, Still Picture Branch, U.S. Weather Bureau, RG 27, Series S, Item 2.

Acknowledgments

Part of this research was supported by grant number 1R01 HD33554 from the National Institute of Child Health and Human Development. The author thanks Myron Gutmann for his support and for his intellectual contributions to this study. The GIS analysis was done on a Dell Inspiron™ 5000 laptop with 128 megabytes RAM using ArcView 3.2. Interpolation of weather data from station points to county polygons was done in ARC/INFO®, as was interpolation of soil data from soil unit polygons to county polygons.

Data sources

Climate data
T. R. Karl, C. N. Williams, Jr., F. T. Quinlan, and T. A. Boden, *United States Historical Climatology Network (HCN) Serial Temperature and Precipitation Data,* Environmental Science Division, Publication no. 3404, Carbon Dioxide Information and Analysis Center, Oak Ridge National Laboratory, Oak Ridge, Tenn. The historical climatology data is stored as point data for weather stations at monthly intervals for 1,221 stations in the United States.

National Climatic Data Center, Arizona State University, and Oak Ridge National Laboratory, Global Historical Climatology Network (GHCN). Comprehensive monthly global surface baseline climate data.

The Great Plains Population and Environment Project (www.icpsr.umich.edu/plains) interpolated data from 394 weather stations in the grassland to counties for each month between 1896 and 1993 by generating a triangulated irregular network (TIN) with ARC/INFO GIS software. That created 3,492 surfaces that were then converted to five-kilometer cell data. The cell data was spatially averaged across each county using a zonalmean function.

Figure 6 maps five-year precipitation deficits for each county. The maps show county precipitation for the five years prior to each dust season as a percentage of average precipitation.

Soil composition
U.S. Department of Agriculture, Natural Resources Conservation Service, State Soil Geographic (STATSGO) Data Base. This is a map database of soil characteristics at a scale of 1:250,000. STATSGO represents soils as polygon mapping units. For each mapping unit there are one to thirteen associated soil types. From the soil type variable (e.g., silty clay loam) a percent of sand, silt, and clay was assigned from a soil texture table. A percent of sand, silt, and clay for each map unit was calculated using a weighted average of its constituent soil types. These mapping units were converted to one-kilometer cell data, then spatially averaged across each county using a zonalmean function.

Agricultural censuses
1925, 1930, 1935

Great Plains Population and Environment Database, 2000, Inter-University Consortium for Political and Social Research (ICPSR), University of Michigan. See Myron P. Gutmann, Sara M. Pullum, Geoff Cunfer, and Delia Hagen, *Great Plains Population and Environment Database Version 1.0 User's Guide.* The updated version 2.0 user's guide is available at www.icpsr.umich.edu/plains.

Map data
Carville Earle, *U.S. Historical County Boundary Files, 1790–1970.* Contact the Department of Anthropology and Geography, Louisiana State University, Baton Rouge. See Myron P. Gutmann and Christie G. Sample, "Sources for the Digital Cartography of the United States," in Michael Goerke, ed., *Coordinates for Historical Maps,* 190–200 (St. Katherinen: Scripta Mercaturae Verlag, 1994).

Dust region maps digitized by the Great Plains Population and Environment Project from Donald Worster, *Dust Bowl,* 30.

Further reading

Bonnifield, Mathew P. *The Dust Bowl: Men, Dirt, and Depression.* Albuquerque, N.M.: University of New Mexico Press, 1979.

The Future of the Great Plains: Report of the Great Plains Committee. Washington, D.C.: U.S. Government Printing Office, 1936.

Great Plains Quarterly 6 (Spring 1986). Special issue on the Dust Bowl.

Gutmann, Myron P., and Geoff Cunfer. "A New Look at the Causes of the Dust Bowl." Charles L. Wood Agricultural History Lecture Series, no. 99-1. Lubbock, Tex.: International Center for Arid and Semiarid Land Studies, Texas Tech University, 1999.

Hurt, R. Douglas. *The Dust Bowl: An Agricultural and Social History.* Chicago: Nelson-Hall, 1981.

Malin, James C. "Dust Storms, 1850–1900." *Kansas Historical Quarterly* 14 (May, August, November 1946): 129–44, 265–96, 391–413.

———. *Essays on Historiography.* Ann Arbor, Mich.: Edwards Brothers, Inc., 1946.

———. *The Grassland of North America: Prolegomena to Its History.* Lawrence, Kans., privately printed, 1961.

Riney-Kehrberg, Pamela. *Rooted in Dust: Surviving Drought and Depression in Southwestern Kansas.* Lawrence, Kans.: University Press of Kansas, 1994.

Wooten, H. H. *The Land Utilization Program, 1934 to 1964: Origin, Development, and Present Status.* USDA Economic Research Service Agricultural Economic Report no. 85. Washington, D.C.: U.S. Government Printing Office, 1965.

Worster, Donald. *Dust Bowl: The Southern Plains in the 1930s.* New York: Oxford University Press, 1979.

Wunder, John R., Frances W. Kaye, and Vernon Carstensen. *Americans View Their Dust Bowl Experience.* Niwot, Colo.: University Press of Colorado, 1999.

Notes

1. Geoff Cunfer, "Common Ground: The American Grassland, 1870–1970," Ph.D. dissertation, University of Texas (1999): 234–304.
2. See, for example, the John L. McCarty Papers, Mary E. Bivens Library, Amarillo, Texas.
3. See William Cronon, "A Place for Stories: Nature, History, and Narrative," *Journal of American History* 78 (March 1992): 1,347–76.
4. Identical maps of dust storm locations in the 1930s were published in Donald Worster, *Dust Bowl,* 28–30, and in R. Douglas Hurt, *The Dust Bowl,* after p. 86. They were digitized from Worster's book for this study.
5. Sources for this account of the physics of wind erosion are Dale A. Gillette, "Production of Dust That May Be Carried Great Distances," in Troy L. Pewe, ed., *Desert Dust: Origin, Characteristics, and Effects on Man,* Special Paper 186 (Boulder, Colo.: The Geological Society of America, 1981), 11–26; Dale A. Gillette and Theodore R. Walker, "Characteristics of Airborne Particles Produced by Wind Erosion of Sandy Soil, High Plains of West Texas," *Soil Science* 123 (1977): 97–110; A. P. Bocharov, *A Description of Devices Used in the Study of Wind Erosion of Soils,* Subhash C. Dhamija, trans., V. Pandit, ed. (Calcutta: Oxonian Press, 1984); Nyle C. Brady and Ray R. Weil, *The Nature and Properties of Soils,* 11th ed. (Upper Saddle River, N.J.: Prentice Hall, 1996), 588–92; N. P. Woodruff and F. H. Siddoway, "A Wind Erosion Equation," *Soil Science Society of America Proceedings* 29 (1965): 602–08; M. Scott Argabright, "Evolution in Use and Development of the Wind Erosion Equation," *Journal of Soil and Water Conservation* 46 (1991): 104–05; F. Bisal and J. Hsieh, "Influence of Moisture on Erodibility of Soil by Wind," *Soil Science* 102 (1966): 143–46; W. S. Chepil, F. H. Siddoway, and D. V. Armbrust, "Climatic Index of Wind Erosion Conditions in the Great Plains," *Soil Science Society of America Proceedings* 27 (1963): 449–52; L. J. Hagen and N. P. Woodruff, "Air Pollution from Dust Storms in the Great Plains," *Atmospheric Environment* 7 (1973): 323–32; Bruce E. Lyles and Leon Allison, "Wind Erosion: The Protective Role of Simulated Standing Stubble," *Transactions of the American Society of Agricultural Engineers* 19 (1976): 61–64.
6. Myron P. Gutmann and Geoff Cunfer, "A New Look at the Causes of the Dust Bowl," Charles L. Wood Agricultural History Lecture Series, no. 99-1 (Lubbock, Tex.: International Center for Arid and Semiarid Land Studies, Texas Tech University, 1999).
7. See the VEMAP weather data project at www.cgd.ucar.edu:80/vemap; this data will soon be interpolated to the county level for the Great Plains states by the Great Plains Population and Environment Project (www.icpsr.umich.edu/plains).
8. Gutmann and Cunfer, "A New Look," 12–20.
9. Jack L. Hofman, Robert L. Brooks, Joe S. Hays, Douglas W. Owsley, Richard L. Jantz, Murray K. Marks, and Mary H. Manhein, *From Clovis to Comanchero: Archeological Overview of the Southern Great Plains,* Arkansas Archeological Survey Research Series no. 35 (Fayetteville, Ark.: Arkansas Archeological Survey, 1989); Eileen Johnson, ed., *Lubbock Lake Landmark: 1987 Fenceline Corridor Survey and Testing Program,* Lubbock Lake Landmark Quaternary Research Center Series no. 1 (Lubbock, Tex.: Museum of Texas Tech University, 1989).

CHAPTER EIGHT

AGRICULTURAL HISTORY WITH GIS

Alastair W. Pearson and Peter Collier

If geographic information systems had been invented by the time Victoria ascended the throne, you can be sure they would have been used for the Tithe Survey of England and Wales. Conducted between 1836 and 1850, the Tithe Survey is a comprehensive inventory of agricultural land in more than eleven thousand parishes. The survey's detailed maps of landholdings are cross-referenced to tables that list each field, its owner, its tenant, and how it was used—ideal for conversion into GIS. Because the Tithe Survey records so much information about agriculture, historians have used it extensively to study what is often called the British agricultural revolution. Few, however, have analyzed Tithe data in relation to other sources. We are using GIS to combine the geographic and economic information contained in the Tithe Survey with other socioeconomic and environmental data for parishes in Britain's agricultural heartland and in southwest Wales.

The view that British agriculture was revolutionized between 1750 and 1850 is based mainly on the dramatic improvements in farm productivity in certain parts of England, most notably in lowland wheat-growing regions such as East Anglia.

Studies of agricultural change in England and Wales credit wealthy landowners with leading the movement for reform. These improvers experimented with new crops, promoted crop rotation and the use of manure and lime to enrich the soil, and sought to increase efficiency and productivity by investing in better machinery and consolidating their landholdings. Scholars also point to the growth of urban markets as a powerful spur to increase productivity and to expand agriculture beyond its traditional locations. Lastly, the replacement of customary rights of tenure with tenancy agreements on fixed terms of years gave tenant farmers new incentives to increase production while husbanding the soil and taking better care of the properties they rented.

We know little, however, about the extent to which these reforms penetrated areas of Britain that were still poorly connected to the market economy in the early nineteenth century. One such area was southwest Wales. The parish of Newport, located on the northern coast of Pembrokeshire, was typical of the region *(figure 1)*. When Newport was surveyed in 1845, it was a small market and fishing village. The nearest town of any size, Swansea, was 60 miles away over rugged hills, or 110 miles by sea. More than 40 percent of Newport parish lies on the rugged upland of Carn Ingli, whose western flank is exposed to strong, salt-laden winds blowing in from Cardigan Bay *(figure 2)*. Acidic parent rock and high rainfall limit soil fertility throughout the parish. Soils are thin. These conditions posed serious problems for farming in the nineteenth century. John Johnes, the assistant Tithe commissioner for Newport, described conditions in the parish in a letter dated 2 February 1844. He wrote that, in the small fields ringing Carn Ingli, "boulders are now to be seen in many instances through the corn, that

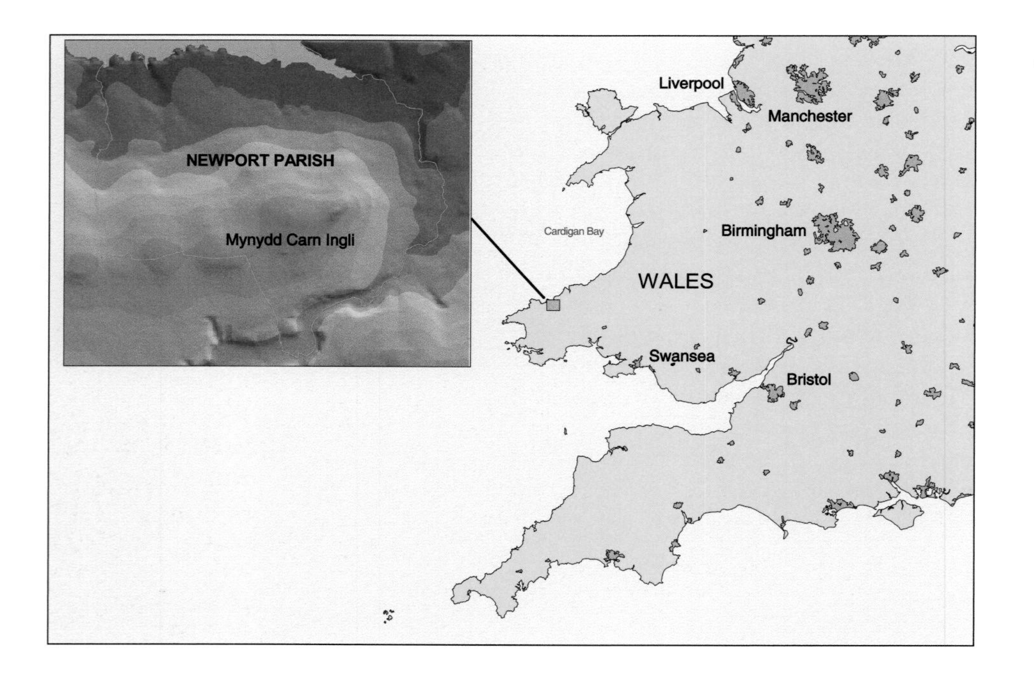

Figure 1. Newport parish
Newport's location, remote from the main urban centers of England and Wales, limited farmers' access to markets for their produce. Some trade went by sea. Drovers took livestock on the hoof across the barren wastes of south Wales to markets in the Midlands and southern England.

they grow potatoes, barley-oats generally, that oats are put in as a first crop sometimes, they manure for potatoes or peat and burn, that the cultivation is chiefly manual on the common on account of the large stones."[1] He mentioned no growing of wheat, the staple cash crop of lowland England. Only potato patches benefited from manure, and some farmers still used the primitive method of burning cropland to return nutrients to the soil. Squatters on the common used handheld implements, not horse-powered machines.

According to local histories, some of Newport's wealthy landlords were implementing agricultural reforms as early as the 1780s. Thomas Lloyd and George Bowen consolidated their holdings for greater efficiency, promoted turnpike trusts to improve transportation, and enclosed unused land for cultivation or raising sheep. Bowen, one of the leading improvers in southwest Wales in the late eighteenth century, was particularly keen to introduce manuring. His experimental mixture of seaweed, marl, and bone meal achieved "most beneficial results."[2]

Other farmers, Welsh historians surmise, were probably aware of innovations but did not implement them. Root crops

Figure 2. Carn Ingli
The rugged, heather- and bracken-strewn flanks of Carn Ingli rise above a relatively fertile coastal plain and river valleys. Archaeological remains and palynological analysis in the area testify to the ebb and flow of marginal cultivation here throughout the last three thousand years. The northeastern rampart of the Iron Age hillfort of Carn Ingli can be seen in the foreground. Photograph by Alastair Pearson.

such as turnips, a favorite among reformers, did poorly in the region's thin, damp, acidic soils. Farmers worried that the turnip fly would destroy the crop and erase their slim capital. The four-course rotation of crops recommended by improvers reduced by half the amount of arable land devoted to food crops. Sympathetic scholars have concluded that Welsh farmers' lack of capital, as much as their suspicion of change, kept them farming in traditional ways as late as the 1840s.

To what extent did landlords and tenants in Newport parish embrace or resist agricultural reforms? To answer this question we turned to the Tithe Survey's wealth of detail about the agricultural landscape. The survey was intended to rationalize the national system for financing the Church of England. Originally amounting to one-tenth of a farmer's produce, the tithe had fallen into disrepute after centuries of abuse and irregular application. "By the beginning of the 19th century," write Tithe Survey historians Roger J. P. Kain and Richard R. Oliver, "it was no longer possible to discern even the vaguest outlines of a system amongst a host of local practices and customary arrangements."[3] The Tithe Commutation Act of 1836 replaced traditional payments with a money rent that fluctuated year to year according to the price of grain. The initial level of rent was determined by surveyors who assessed the kind and value of land on every farm. They assigned each parcel of land to a descriptive category and gave it a monetary value that represented its productive potential.

The lists of land parcels and valuations that the surveyors prepared, called the Tithe apportionments, also name the owner and tenant (if any) of each farm *(figure 3)*. This makes it possible to link land use to other contemporary socioeconomic data that is organized by household. The 1841 Census of Population was conducted just four years before the Tithe Survey came to Newport. The census lists every household in the parish under the head of household, whose name in most cases appears in the 1845 survey as either the tenant or freehold owner of a farm or dwelling. The census further lists all members of the household, their age, sex, and occupation. In addition, the geographical detail of the Tithe maps makes it feasible to compile their field boundaries with detailed environmental data from present-day digital sources.

We have brought together Tithe, census, and environmental data in a GIS. Field boundaries in Newport have changed little since 1845, so to create the basemap

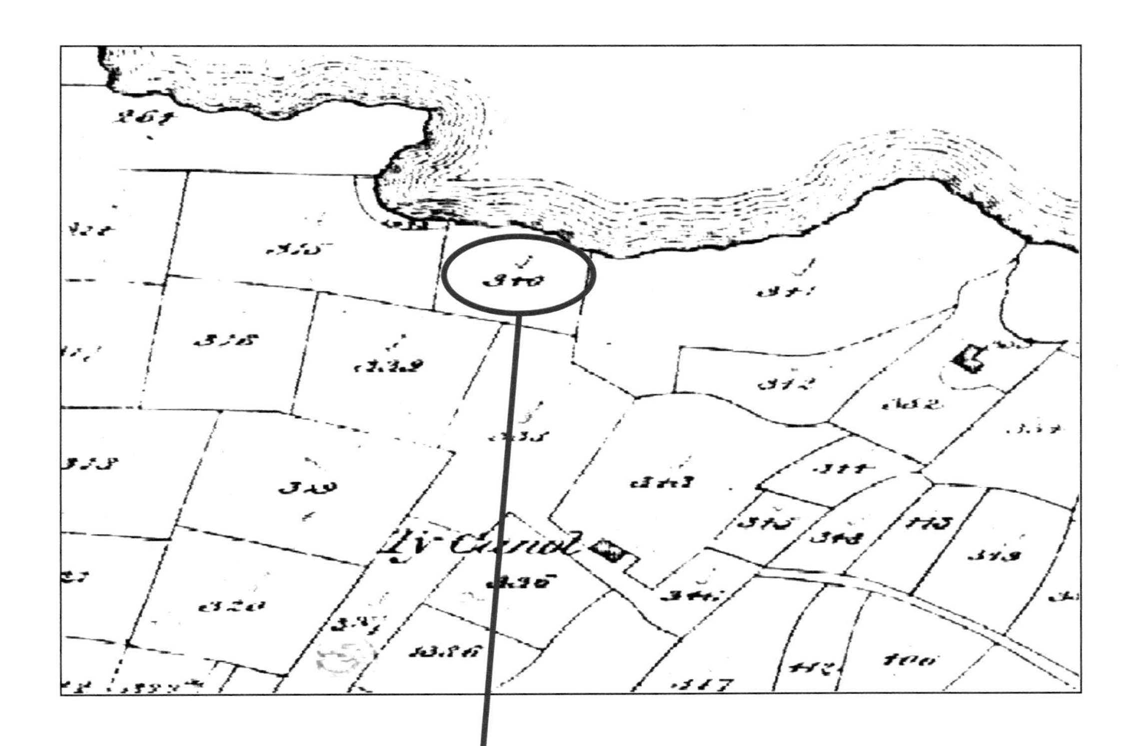

LANDOWNERS.	OCCUPIERS.	Numbers referring to the Plan.	NAME AND DESCRIPTION OF LANDS AND PREMISES.	STATE OF CULTIVATION.	QUANTITIES IN STATUTE MEASURE. A.	R.	P.	Amount of Rent-Charge apportioned upon the several Lands, and payable to RECTOR. £.	s.	d.
				Brought forward	48	2	26	5	12	6
Bowen George (Continued)	Bowen Benjamin (Continued)	287	Field	Pasture		2	17		1	7
		288	"	Pasture		1	6			11
		289	"	Pasture		3	20		2	6
		290	"	Pasture		3	26		2	6
					51	1	9	6	.	.
	Griffiths David	414	Field	Pasture		3	11		3	2
		415	"	Arable	3	3	36		13	1
		416	"	[illegible]	1	1	8		5	.
		417	"	Arable		3	23		3	9
		418	"	Pasture		2	24		2	6
					7	2	27	1	7	6
	Evans James	337	Plain [illegible]	Arable	2	1	18		5	8
		338	Park y West	Arable	3	1	18		5	.
		339	Park [illegible]	Arable	2	2	4		4	4
		340	Park bach [illegible]	Arable	1	1	22		1	11
		341	[illegible]	Arable	6	3	3		5	7
		342	Field	Pasture	1	2	34		2	6
		343	House, Garden & field	Pasture	4	.	14		17	6
		344	Hill	Arable		2	29		1	.
		345	"	Arable		2	36		1	.
		346	"	Arable		3	14		1	6
		347	Park [illegible]	Arable	2	2	5		7	2

Figure 3. A Victorian Land Information System
This extract from the Newport Tithe map and apportionment shows the simple numerical system that links field numbers on the map to the list of fields, by farm, in the apportionment. 1845 Newport Tithe map and Tithe schedule details courtesy of the National Library of Wales.

we digitized the latest large-scale Ordnance Survey maps (at a scale of 1:10,000) and edited them to match the Tithe map.[4] All spatial data was input using the same coordinate system. The mean altitude, slope, and aspect for each field were derived from digital elevation data supplied by the Ordnance Survey. Soil characteristics were provided by the Soil Survey of Great Britain. We also entered the 1841 census data for the head of household of each property listed in the Tithe Survey.

Figure 4 and table 1 show land use in Newport parish, as classified in the "state of cultivation" column of the Tithe apportionment. The broad definitions of land-use categories in the instructions to Tithe surveyors, and the variations in how surveyors interpreted those definitions, make it difficult to draw definitive conclusions about land use. We do know, however, that Henry Phelps Goode, the Tithe surveyor and valuer for Newport, was familiar with the area and that local farmers did not dispute his valuations. The pattern of land use Goode recorded was typical of areas with mixed farming in early nineteenth-century Wales. Just over 17 percent of all agricultural land in Newport was arable, while 63 percent was

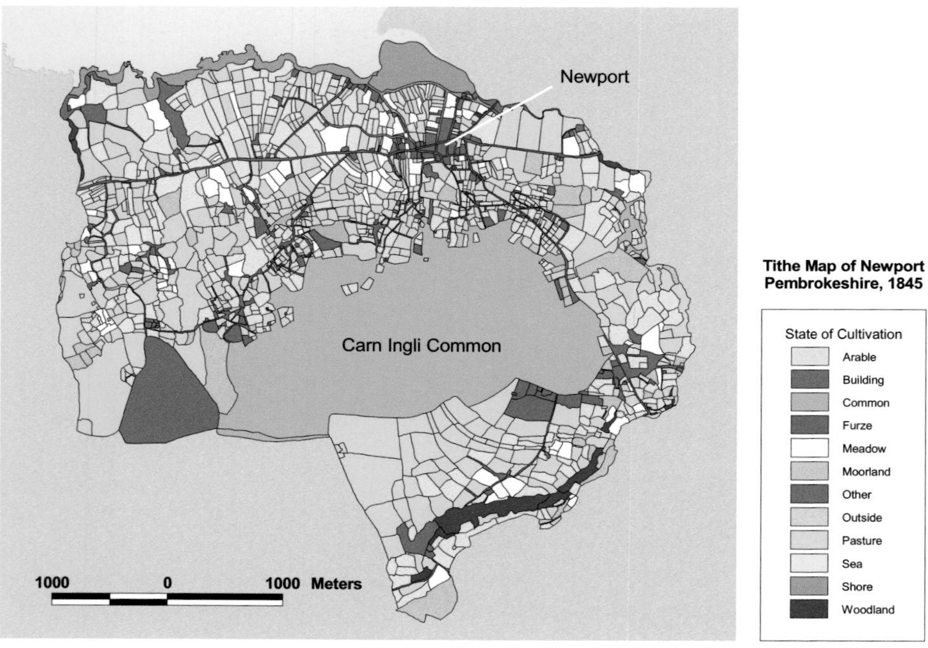

Figure 4. State of cultivation as recorded by the Tithe Survey
Note how arable (cropland) dominates land use along the coastal plain, while large tracts of pasture lie to the south of the Fishguard-to-Newport Road. Small fields of intermixed pasture and arable border the northern edge of the common.

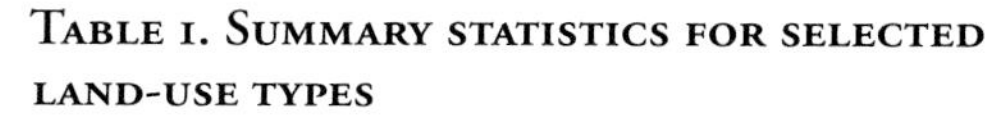

Table 1. Summary statistics for selected land-use types

Land-use type	Fields	Acreage	% land use
Arable	496	797.75	17.32
Common	7	1,223.74	26.57
Cottage	30	28.69	0.62
Furze	19	45.03	0.98
Garden	21	4.16	0.09
House	109	64.40	1.40
Meadow	89	210.57	4.57
Moory pasture	51	202.20	4.39
Pasture	474	1,117.57	24.26
Pasture/Furze	23	309.61	6.72
Wood	12	84.36	1.83
Other (nonfarm)			11.25

some kind of pasture, including the large common on Carn Ingli. This was a community that still practiced transhumance on a local scale, with many farmers pasturing livestock on the common in the summer and feeding animals hay from small meadows at lower elevations in the winter. As elsewhere in Pembrokeshire at this time, woodland was sparse; it was concentrated along the banks of the River Gwaun on the southern edge of the parish.

A handful of men and women owned most of the land in Newport *(figure 5 and table 2).* Thirteen of the seventy-three listed by the apportionment owned 93 percent of the land, and two men—the improvers Thomas Lloyd and George Bowen—owned 61 percent. The other sixty landowners possessed small holdings, some less than half an acre. Although most of the parish's land was under cultivation or used for grazing livestock, 31 percent of Newport's working people were engaged in occupations connected with the sea, whether the coastal trade of agricultural produce, slates from local quarries, or fishing. Only 5 percent of the working population listed "farmer" as their occupation, and another 10 percent called themselves farm laborers. Seven

Table 2. Number of fields, total acreage, and percentage of land owned by those owning more than 1 percent

Owner	Fields	Acreage	% land use
George Bowen	445	1,259	27.3
Common (Thomas Lloyd)	8	1,223	26.5
Thomas Lloyd	372	335	7.3
Esther Bowen	74	249	5.4
Rev. Daniel Davies	29	239	5.1
Lady Mathias	47	185	4.0
Rev. Peter Richardson	24	154	3.3
David Hughes	39	151	3.2
Richard Lobant	42	143	3.1
William Harries	65	81	1.8
Benjamin Evans	41	77	1.7
Trustees of Llangloffan Chapel	23	76	1.7
Rev. Llew. Lloyd Thomas	31	54	1.2
William Morgan	29	53	1.2
Others	331	327	7.1
	1,569	**4,606**	

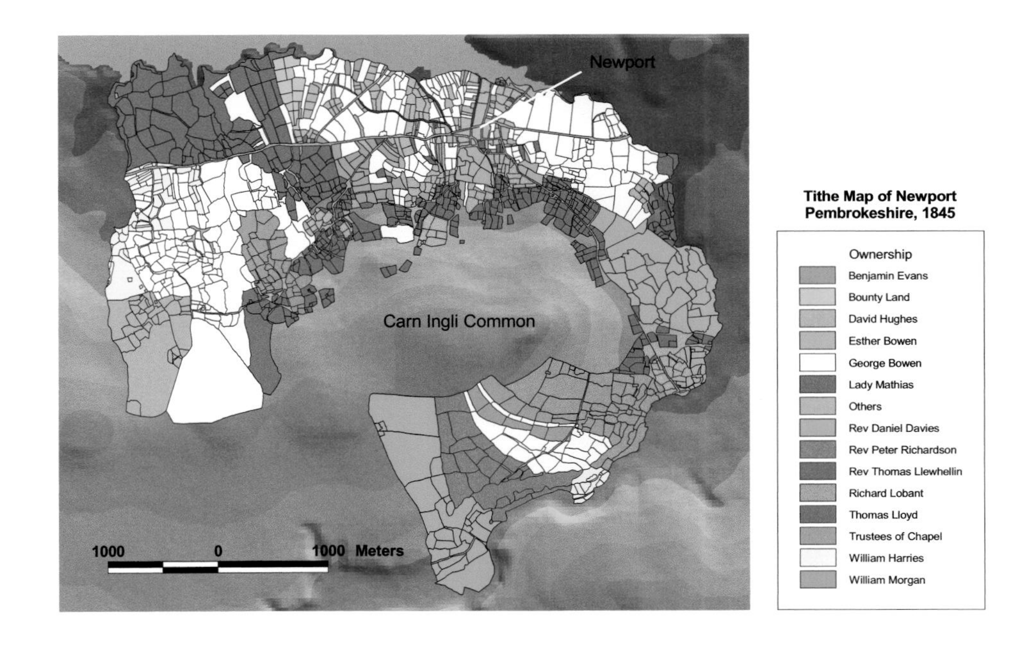

Figure 5. Landownership as recorded by the Tithe Survey
Most land was owned by a handful of owners. George Bowen owns the largest share. Thomas Lloyd, Lord of Cemais, appears to own land fringing the common. However, this land was occupied by squatters who contested his ownership rights.

percent had retail occupations such as milling, baking, and shopkeeping, and 10 percent worked in the domestic textile industry as cottage spinners, weavers, or knitters. As in many other places across southwest Wales, many people in Newport pieced together a patchwork subsistence that combined nonfarming occupations with keeping a garden and potato patch, raising a little livestock, and earning wages as farm laborers during harvest time. Smallholders' ties to agriculture might have been more attenuated still had they not owned relatively fertile land *(figure 6)*.

Figure 6. Tithe rent per acre
The geographical pattern of Tithe charges is clearly not random and demonstrates that the rent charge was apportioned consistently. Rent charge decreases as one moves away from Newport and onto higher elevations with thinner, stonier soil.

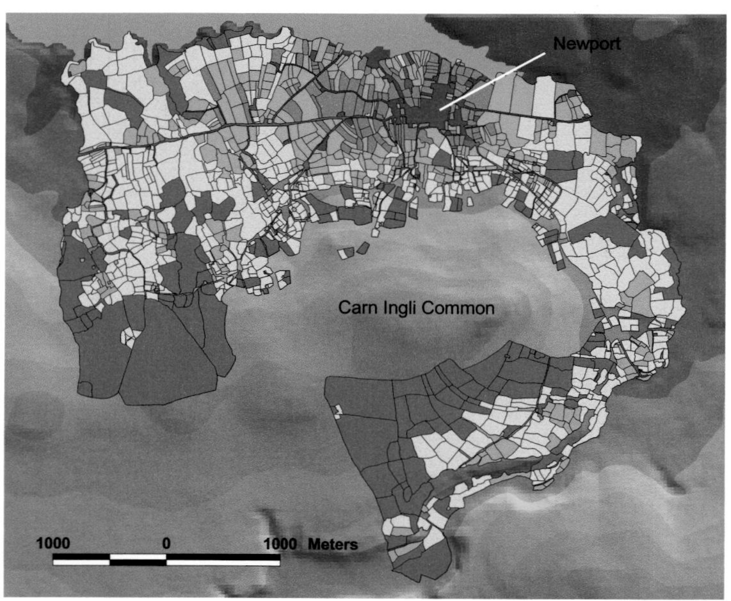

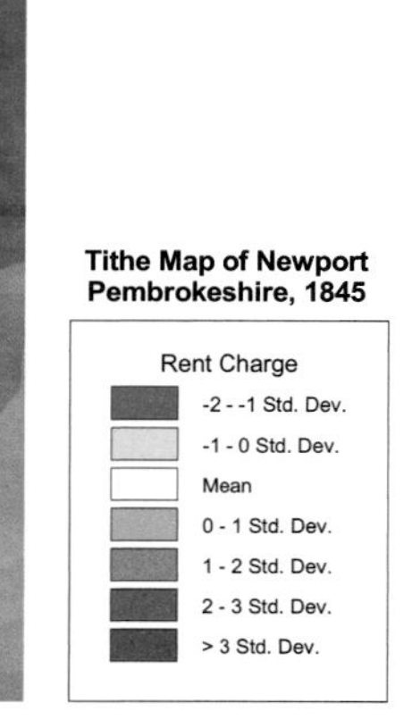

The pattern of rents follows local topography closely. No tithe was levied on the rough, steep land of Carn Ingli common. Most of the lowest rents applied to rough pasture surrounding the common. Where altitude drops to the west and south of Carn Ingli, there were pockets of higher rent. The highest rents per acre were clustered around Newport, along the coast, on gently rolling ground. This was where most smallholdings were located. It was also where farmers had easiest access to seaweed to fertilize their fields and to roads for hauling manure and lime. Living close to the village of Newport made it easier for smallhold farmers to have many jobs.

The geographical patterns of land value and land use suggest that topography imposed limits on agriculture in Newport parish, but do not tell us the significance of particular environmental factors or the varying effects on productivity of landowners and occupants.[5] To distinguish the effects of human agency (and whose agency) from the effects of natural endowments, we needed to apply statistical analysis to the data on owners, occupants, and physical characteristics of each field, and to relate those factors to the rents assessed by the Tithe surveyors.

We first used multiple regression[6] to analyze the relative importance of environmental factors on the Tithe rent charged for each field. The rent on 62 percent of fields was primarily constrained by altitude, compared to 16 percent by slope, and just 7 percent by distance from Newport. No field's rent was primarily determined by soil type or aspect. We do not take these percentages as clear or decisive measures because the data on which they are based is not entirely reliable at the scale of this local study. For example, the Soil Survey data for Newport shows that soil type tends to vary with slope and altitude, but does not show precise variations in soil quality from field to field. The way farmers used land also influenced the quality of soil and its productivity. When we added dummy variables for each land use in the multiple regression analysis, the resulting model explained 60 percent of the variation in rents.

In addition to the limitations of the data, multiple regression itself is limited in its capacity to explain geographical variation. Regression models assume random, independent distributions over space; they are not intended to model the clustering and dispersal that are fundamental to geographical variation. Regression models are also not well suited to analyzing relationships between more than two levels of potentially influential factors. To take a different example, educators believe that children's academic performance depends largely on such factors as innate intelligence, family circumstances, and health. But one must also weigh the significance of teachers, administrative systems, and resources. Multilevel modeling[7] is much more effective than regression in analyzing hierarchical structures of this kind. In our study, geographical variation existed at three levels, in the nested relationships of fields to occupants and of occupants to landowners. Having established that Newport's physical environment posed very significant constraints on agricultural productivity, we used multilevel modeling to determine whether rent variation was also influenced significantly by owners and occupants.

The results showed that 79 percent of rent variation was accounted for at the level of the field (the influence of environmental factors), 12 percent at the level of the occupants, and 9 percent at the level of the owner. These findings suggest that occupants and owners did marginally affect productivity. We cannot draw too many conclusions from statistical analysis of such a small study area. However, tenant farmers appear to have been slightly more responsible for agricultural improvement in this part of Pembrokeshire than were landowners. Despite their efforts to institute new methods of husbandry, Bowen and Lloyd may have had less success in improving the productivity of their land in this parish than did smaller farmers and tenants. Our initial findings challenge earlier portrayals of Welsh tenant farmers and smallholders as deeply opposed to new methods of farming.[8] They also confirm historian David W. Howell's conclusion that "it may have been that the minor gentry and those large freeholders and tenant farmers below them did most to advance the techniques of agriculture by dint of patient trial and error."[9]

We will have a better sense of the significance of our findings once we complete similar analyses of parishes in Somerset, Cambridgeshire, Hampshire, and West Sussex. Comparing Newport to these English counties will help us understand variation within the so-called national agricultural revolution much more fully. At this point our diverse methodology, combining GIS with multiple regression and multilevel modeling, promises to shed new light on past landscapes and the economic dynamics they embodied.

This project demonstrates the potential value of GIS to the historian interested in reconstructing past agricultural landscapes. The ability to visualize patterns in both the human and physical landscape and then go on to analyze relationships between them offers enormous potential for future work. Technology now allows us to articulate historical data in a dynamic way, making the exploration of these past landscapes a fascinating avenue of research. We can integrate historical and environmental data, thus offering us the opportunity to employ a diverse methodology that combines GIS with multiple regression and multilevel modeling. New light can be shed on past landscapes and the economic dynamics they embodied. Prior to GIS such possibilities were merely a pipe dream. However, it takes months of work to input the data in a form compatible with the GIS. Indeed, not all documents and manuscripts lend themselves to being input into a GIS. For example, files that accompany the Tithe Survey held by the Public Record Office provide invaluable details on the methods of cultivation and quality of the land, yet offer very poor spatial referencing. The success or failure of the application of GIS to this type of study depends on the willingness of the researcher not to forsake the traditional methods and techniques appropriate to the analysis of a diverse range of sources. Furthermore, it has no inherent answers, only those of the analyst.

Acknowledgments

The authors would like to acknowledge the thoroughness and tireless effort of Anne Knowles during the completion of this paper. We would also like to thank Professor Kelvyn Jones and Dr. Myles Gould for their assistance with the multilevel modeling analysis. Thanks are also due to Paul Carter and Rut Galmeier for their help with the digitizing of the Tithe map and to Karen Pearson for her patience and encouragement throughout.

Further reading

Burrough, P. A., and R. McDonnell. *Principles of Geographical Information Systems for Land Resources Assessment,* 2d ed. Oxford: Clarendon Press, 1998.

ESRI. *Understanding GIS: The ARC/INFO Method,* 4th ed. Cambridge: GeoInformation International, 1997.

Gregory, I. N. *A Place in History: A Guide to Using GIS in Historical Research.* University of Essex: History Data Service, forthcoming.

Howell, D. W. *Pembrokeshire County History.* Vol. 4, *Modern Pembrokeshire 1815–1974.* Haverfordwest: Pembrokeshire Historical Society, 1993.

Jones, K. J. "Specifying and Estimating Multi-level Models for Geographical Research." *Transactions of the Institute of British Geographers* 16 (1991): 148–60.

Kain, R. J. P., and H. C. Prince. *The Tithe Surveys of England and Wales.* Cambridge: Cambridge University Press, 1985.

———. *Tithe Surveys for Historians.* Chichester: Phillimore, 2000.

Longley, P. A., M. F. Goodchild, D. J. Maguire, and D. W. Rhind, eds. *Geographical Information Systems: Principles, Techniques, Management and Applications,* 2d ed. Chichester: John Wiley, 1999.

Pearson, A. W., and P. Collier. "The Integration and Analysis of Historical and Environmental Data Using a Geographical Information System: Landownership and Agricultural Productivity in Pembrokeshire c. 1850." *Agricultural History Review* 46 (1998): 162–76.

Shepherd, I. D. H. "Information Integration and GIS." In D. J. Maguire, M. F. Goodchild, and D. W. Rhind, eds., *Geographical Information Systems: Principles and Applications,* Vol. 1, *Principles,* 337–60. Harlow: Longman (1991). Available online at www.wiley.co.uk/wileychi/gis/resources.html.

Unwin, D. "Integration through Overlay Analysis." In M. Fischer, H. J. Scholten, and D. Unwin, eds., *Spatial Analytical Perspectives on GIS,* 129–38. London: Taylor & Francis, 1996.

Notes

1. Public Record Office, IR18/14707.
2. F. Jones, "Bowen of Pentre Ifan and Llwyngwair," *Pembrokeshire Historian* 6 (1979): 25–57.
3. Roger J. P. Kain and Richard R. Oliver, *The Tithe Maps of England and Wales: A Cartographic Analysis and County-by-County Catalogue* (Cambridge: Cambridge University Press, 1995), 1.
4. Instead of digitizing printed maps, we now edit current Ordnance Survey digital Landline data derived from 1:2,500 mapping of rural areas. Digitizing of the Tithe map is used only when there has been significant change.
5. The Tithe Survey apportionment categorizes individuals as "landowners" and "occupiers." Occupiers, or occupants, include large and small landowners as well as tenant farmers. The category meant whoever was working a particular farm or plot of land at the time of the survey.
6. Multiple regression is a parametric statistical technique for identifying the relationship between a dependent variable (here, Tithe rent) and one or more independent variables (land use, soil, altitude, and so forth). The technique fits a straight-line plane to the trend in a scatter of points.
7. Multilevel modeling recognizes hierarchical structures in the data and permits one to assess the influence of explanatory variables on each of those levels simultaneously.
8. Sir Charles Hassall, *General View of the Agriculture of the County of Pembroke with Observations on the Means of Its Improvement* (London: Smeeton, 1794).
9. David W. Howell, "The Economy, 1660–1793," in B. E. Howells, ed., *Pembrokeshire County History,* Vol. 3, *Early Modern Pembrokeshire,* 315 (Haverfordwest: Pembrokeshire Historical Society, 1987).

CHAPTER NINE

MAPPING BRITISH POPULATION HISTORY

Ian N. Gregory and Humphrey R. Southall

Vital statistics are the lifeblood of modern government bureaucracy. They are also the numbers that have enabled quantitative social historians to recover the lives of ordinary people and to study profound changes in population history. European nations began to keep systematic records of the social characteristics and economic activities of their populations around the turn of the nineteenth century. The first British census, for example, was conducted in 1801. Regular censuses provided the basis for new methods of government regulation, planning, and control. They reflected a general fascination with statistics and keeping strict accounts that also surfaced in the widespread adoption of double-entry bookkeeping and the detailed membership records kept by many religious denominations. As government officials discovered how useful statistical information could be, they enlarged the census to collect more kinds of information and began to gather statistics on elections, social conditions, migration, welfare payments, and other matters related to policy and governance.

Keeping track of people necessarily means knowing where they are. That is why governments usually gather and publish social statistics for defined territorial units. The problem is that the definition of units has changed over time. The British government fundamentally redefined the national system of "districts" (the main unit of local government) three times between 1840 and 1974. The changes came in response to shifting political power, liberals favoring expanding the number of districts to give localities greater voice, conservatives wanting fewer, larger districts to centralize control. From 1840 through World War I, local administration was handled by approximately 635 registration districts, most of which corresponded closely to the Poor Law unions on which they were based. After the war, registration districts were replaced by about fifteen hundred local government districts. In 1974, the government redesigned the system more radically, reducing the total number to just 330 districts, most of which were much larger than their predecessors.

In addition to the wholesale remapping of governmental units, officials have made thousands of smaller, incremental changes to unit boundaries to maintain a rough equality in the numbers represented by each district as cities grew and rural population declined. Figure 1 shows the changing administrative boundaries of the county of Gloucestershire between 1851 and 1971. Urban districts multiplied between 1911 and 1931, reflecting the growth of Bristol, Cheltenham, and Gloucester. From 1931 to 1971, rural districts in the north of the county were consolidated. There were also smaller boundary changes. Between 1851 and 1911, for example, the Bristol registration district expanded greatly, absorbing sections of a number of outlying districts.

The precise geography of administrative unit boundaries may seem a trivial thing. Few British population historians have acknowledged that the boundaries of British census units have changed thousands of times since 1801. In fact, administrative boundaries of all kinds have changed throughout British history.[1] Failing to account for the relocation of unit boundaries can seriously undermine the reliability of statistical analysis, particularly when the results of analysis are presented cartographically. To map and spatially analyze data correctly, quantities must be linked to an accurate representation of the units for which they were collected. And to compare quantities collected for territorial units over time, we must somehow acknowledge how unit areas changed, either by explicitly noting the inaccuracy of direct comparison or by statistically removing the impact of boundary changes.

Unfortunately, it is difficult to find accurate boundaries for British population censuses. The General Register Office (GRO) has published only a few, poor-quality boundary maps for England and Wales over the past two hundred years. Faced with this obstacle to obtaining accurate boundaries, many historians do not map their findings, denying themselves and their readers what is often the most effective means of identifying patterns of local, regional, and national population patterns. Historians who do plot socioeconomic statistics resort to the census map published nearest to their study date or, in the most egregious cases, simply use boundaries from the most convenient map they can find, whatever its date and source.

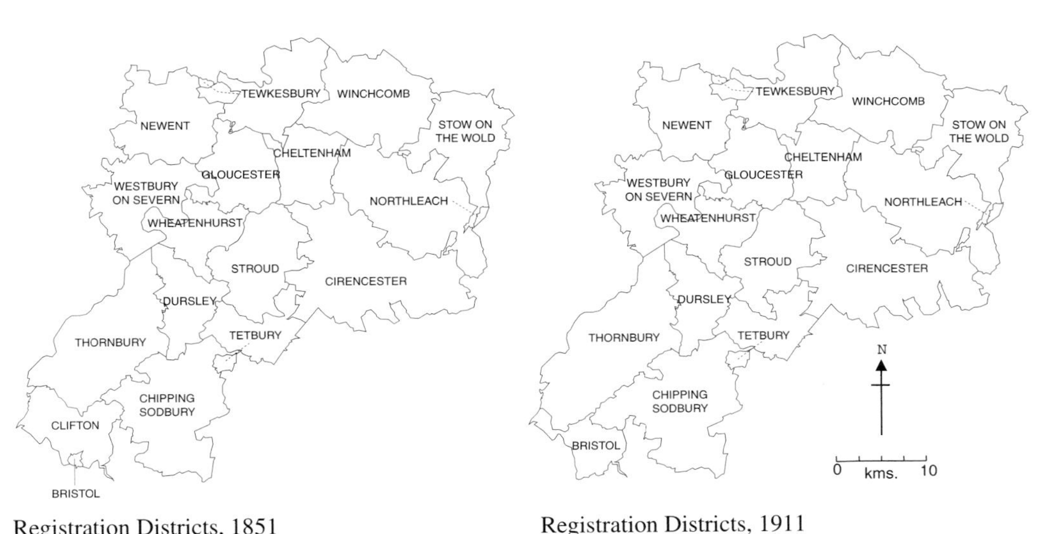

Registration Districts, 1851

Registration Districts, 1911

Figure 1. Administrative units in Gloucestershire, 1851–1971

These maps show how one county's administrative geography changed over 120 years. Note the smaller urban districts added by the change from registration districts to local government districts between 1911 and 1931, and the continuing expansion of Bristol. Dotted lines link detached portions to their main district.

Local Government Districts, 1931

Local Government Districts, 1971

County Boroughs
1. Gloucester, 2. Bristol

Municipal Boroughs
3. Tewkesbury, 4. Cheltenham

Urban Districts
5. Charlton Kings, 6. Stow on the Wold,
7. Coleford, 8. Westbury on Severn,
9. Newham, 10. Awre, 11. Stroud,
12. Cirencester, 13. Nailsworth, 14. Tetbury,
15. Mangotsfield, 16. Kingswood

Named areas for 1931 and 1971 are all Rural Districts

The Great Britain Historical GIS Project was organized in 1994 to bring geographical accuracy to the study and representation of population history. The goal of the project was to build a flexible net of administrative unit boundaries that could capture the fluid reality of society as accurately as possible for any date and for any of several kinds of units, giving social science historians a better alternative to the rigid, often anachronistic frames available in printed maps. The cells of the net would change shape to reflect boundary locations for the date and unit specified. GIS was the ideal instrument for constructing the boundaries net and its statistical catch because the basic architecture of the system consists of the two components the project required: a spatial database (in this case, polygons representing the exact shape and location of administrative units) and the attributes associated with those polygons.

The first step in building the historical GIS was to gather all available information about boundary changes to the key units of local government for the period of the British census, namely Poor Law unions, registration districts, and local government districts. We also decided to map boundary changes to the country's nearly fifteen thousand parishes. These venerable units of ecclesiastical and welfare administration provide the most detailed geographical framework for analyzing social change at the national scale.[2] Their boundaries were subject to enormous change in the late nineteenth and early twentieth centuries. Before that, however, the relative stability of parish boundaries over centuries enables scholars to make rough but meaningful statistical comparisons between modern Britain and previous eras as far back as the early middle ages.

Most of our digitizing was done from print maps produced by the Ordnance Survey, Britain's national mapping agency, maps that are exceptionally detailed and precise. They display virtually all governmental boundaries as well as information on settlements, lines of communication, and physical features. The earliest series of suitable maps, published in 1888, used a scale of four miles to the inch (1:253,440). Revised maps published after about 1910 were more accurate still, at a scale of two miles to the inch (1:126,720). After World War II, these were gradually replaced by maps at 1:100,000. In addition, we used some nineteenth-century maps at 1:63,360. Ideally one would wish for a single scale across the whole time span of the GIS, but the technology allows us to minimize the differences between scales and to integrate data from various scales. It is important to note that the scale of the source maps does fundamentally limit the uses that the GIS can be put to. The GIS is ideally suited to county- or national-level analyses but is not appropriate for detailed local studies where larger-scale source maps would be required.

Ordnance Survey maps published around 1910 captured the changeover from Poor Law unions and registration districts to local government districts, making that series particularly valuable as a record of sweeping boundary changes. Figure 2 shows an example from the 1909 Ordnance Survey map of Bristol, the largest city in Gloucestershire. Thick red lines show the boundaries of Poor Law unions, such as Bristol and Keynsham. Thin red lines trace parish boundaries. The remaining lines represent other kinds of local government districts.

All boundaries from 1840 to 1974 were digitized and georeferenced to the British National Grid, the standard map projection for Great Britain. This created GIS coverages of the main types of administrative units. The next stage was to enable the system to reproduce boundaries by

Figure 2. Extract from the 1909 Ordnance Survey map of Gloucestershire
The complex layers of administrative units include parishes, bounded by thin red lines, and Poor Law unions, marked by thick red lines and red capital letters. Rural local government districts have thick blue boundaries, while urban districts are shaded pale orange and the county borough of Bristol, pink. The borough is an ancient designation that granted towns special powers, such as the right to hold a market.

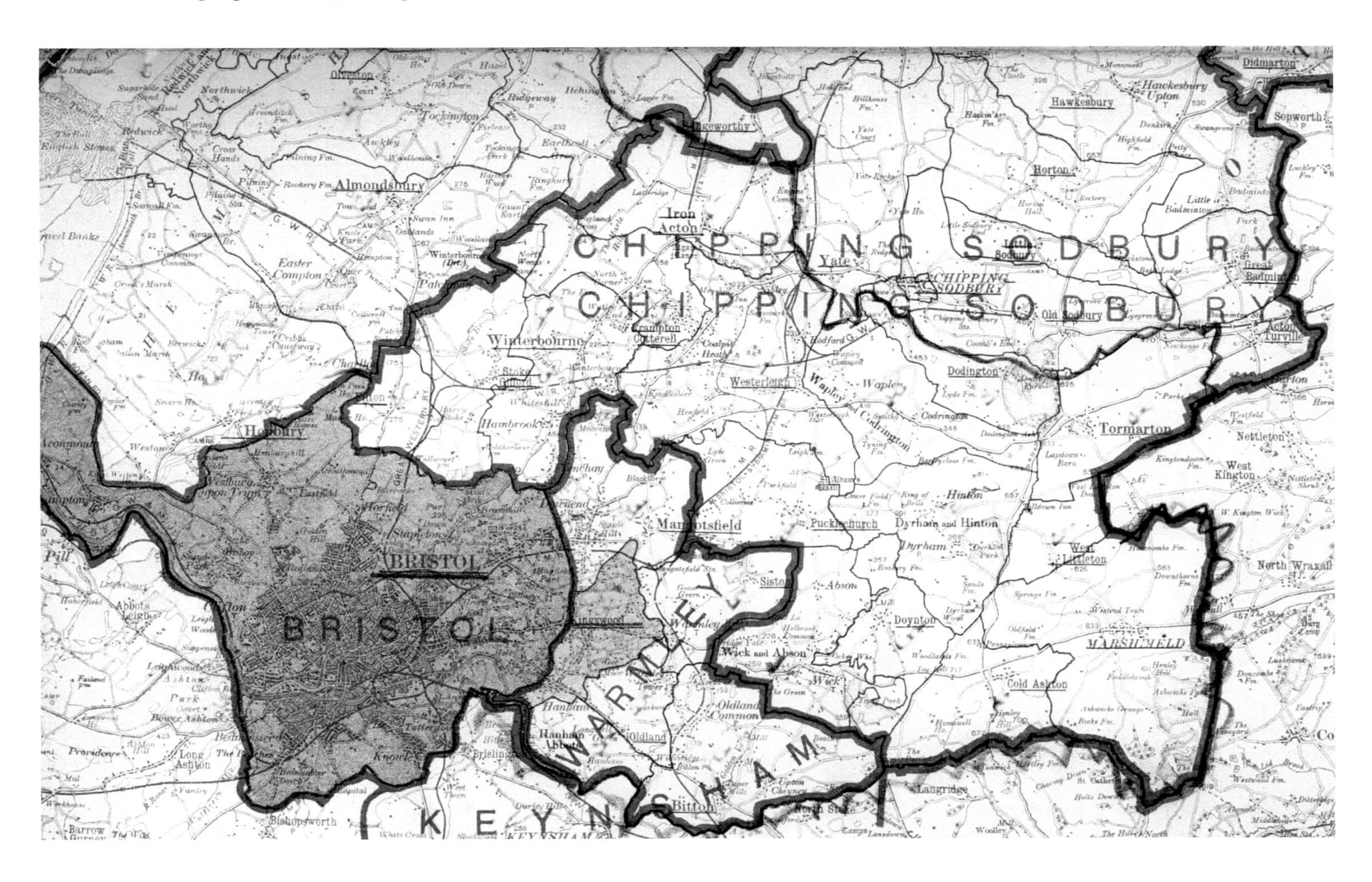

date. To do this, we needed to know when boundaries changed—information provided in tabular form by the Government Records Office and the Local Government Board *(figure 3)*. To bring together the tabular and map-based information, we printed copies of the circa 1910 boundaries onto tracing paper and then drew every boundary change listed in the tables, referring to Ordnance Survey and other historical maps for the location of each boundary. The last step was to enter all the data into the GIS by digitizing boundary lines and tagging each line with its start and end date.

Seven years of research and GIS construction produced a continuous record of the changing administrative geography of Britain. Users enter the unit and date required. At the press of the command button, the GIS displays an outline map built instantly from the time-coded boundary data. To map census data and other socioeconomic statistics, one types a query requesting, for example, statistics on overcrowding for local government districts from the 1931 census *(figure 4)*. The broad patterns in figure 4 are quite predictable: areas with the highest residential density are primarily urban, while the

CHANGES IN REGISTRATION DISTRICTS AND SUB-DISTRICTS IN THE YEAR 1905—*continued.*

Date when change came into operation.	DISTRICT and Sub-District. Decreased by change. Name.	No.	Increased by change. Name.	No.	Name of Parish transferred.	Population, 1901.
1905. 1 Jan.	BARTON REGIS - Westbury on Trym (abolished).	321 1	BRISTOL - - Ashley - - (re-numbered 8 on 1st Dec.)	320 9	Horfield, added to Bristol Civil parish and County Borough.	1435
"	BARTON REGIS - Westbury on Trym (abolished).	321 1	THORNBURY - Almondsbury -	323 1	Henbury - - - - - -	1922
"	BARTON REGIS - Winterbourne- (abolished).	321 2	CHIPPING SODBURY Winterbourne- (new sub-district).	322 1	Filton - - - - - -	464
					Stoke Gifford - - - - -	395
					Winterbourne - - - - -	3624

Figure 3. Boundary change report from Bristol region
This example from the Registrar General's Annual Report, 1905, shows how changes were recorded. The first entry records that on 1 January 1905, the district of Barton Regis lost 321 people, 320 of whom were transferred to Bristol, as a consequence of the transfer of Horfield parish into Bristol Civil parish and County Borough. The parish's transfer affected 1,435 people in all.

lowest averages are found in rural areas. There are surprisingly high rates, however, in the northeast and in coastal districts of the northwest, as well as lower rates near the center of London than one would expect. One of the advantages of using GIS is that maps such as this can be produced as soon as the data is entered, so that the patterns they reveal become integral to the research.

Using historically correct boundaries raises mapping to the same standards of accuracy and responsible use that historians have long applied to quantitative analysis and archival sources. The three maps in figure 5 use historic boundaries to display the proportion of young women in district population as a whole. Looking closely, one can see that the shape and size of districts changes from 1881 to 1911, with even more changes in 1931. In addition to preserving the integrity of the maps' statistical content, using historical boundaries in this case highlights the concentration of young women in urban areas. Seeing patterns shift within changing geographies can help researchers and teachers think about data, raising questions and focusing attention on areas of greatest change or those showing quiescence or decline.

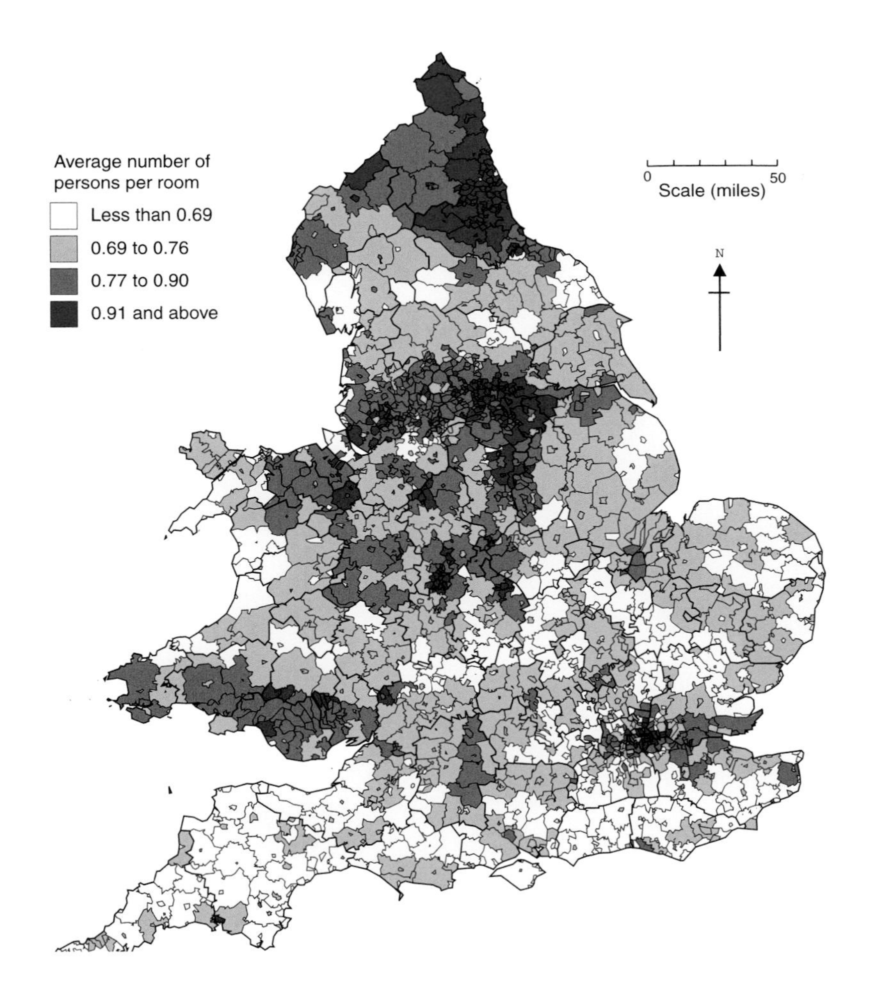

FIGURE 4. OVERCROWDING IN 1931 OVERCROWDED HOUSING WAS A SERIOUS PROBLEM IN ENGLAND AND WALES BETWEEN THE TWO WORLD WARS. ONE WAY THE CENSUS TRIED TO QUANTIFY OVERCROWDING WAS BY CALCULATING THE AVERAGE NUMBER OF PERSONS PER ROOM IN EACH DISTRICT. THE MEAN (AVERAGE) FOR ENGLAND AND WALES WAS 0.76 PER ROOM. DISTRICTS SHADED PURPLE HAD THE HIGHEST RATES AND THE MOST SEVERE OVERCROWDING.

Figure 5. Women aged 15 to 24 as a percentage of total population in Gloucestershire
All three maps show that young women were most concentrated in urban districts; Cheltenham and Bristol had among the highest rates in 1881 and 1911. Although the details are markedly different in the 1931 map (based on local government districts rather than registration districts), the overall pattern remains much the same, suggesting the continuing migration of young women from rural to urban areas.

GIS is about more than simple mapping. It also allows us to analyze quantitative data in ways that are impossible using conventional, aspatial statistics. One of the long-term goals of the Great Britain Historical GIS Project is to be able to study socioeconomic change without having to worry about the effect of boundary changes—to make a GIS capable of generating statistical comparisons that automatically compensate for known changes in boundary locations. This may be easier to grasp by considering table 1, which lists the same categories of data presented in figure 5. On their face, the percentages appear straightforward and reliable. As we now know, however, boundary changes made "Bristol" and other Gloucestershire districts different geographical entities in 1881 than they were in subsequent years. The territory of only three districts in table 1 (Cheltenham, Gloucester, and Westbury on Severn) remained sufficiently stable to support the usual conclusion that population changed by the amount indicated in the table. Boundary changes so altered the other three districts that the table misrepresents the data. It implies that the proportion of young women in a given district changed by a certain amount when in fact their proportion may have remained the

Table 1. Women aged 15 to 24 as a percentage of total population, selected districts in Gloucestershire

District	1881	1911	1931
Cheltenham	12.8	11.5	9.6
Bristol	11.1	10.2	9.1
Barton Regis	11.1	—	—
Stroud	10.2	10.0	8.6
Gloucester	9.8	9.5	8.8
Westbury on Severn	6.9	7.2	6.7

Source: Printed census reports, 1881, 1911, 1931

1881
6.6 to 7.6
7.7 to 8.7
8.8 to 10.2
10.3 to 12.9

1911
6.5 to 7.6
7.7 to 8.4
8.5 to 9.8
9.9 to 11.8

1931
5.9 to 7.1
7.2 to 7.9
8.0 to 8.5
8.6 to 9.6

N
0 kms. 10

All legends use nested means and show the number of females aged 15 to 24 as a percentage of total population.
Source: Printed census reports 1881, 1911, 1931

same or changed in the opposite direction, depending on the nature and magnitude of boundary changes. It certainly was not true that the area called Barton Regis in 1881 had no women aged 15 to 24 in 1911 or 1931. The district, not its population, disappeared.

Accurately calculating the precise statistical effects of boundary changes on population districts is a complicated process. Remapping all census data onto a single set of target districts proved a manageable way to improve the geographical accuracy of longitudinal comparisons. This process, called *areal interpolation,* reallocates data from its original units to the target units. Technically, this is done by overlaying the source units onto the target units and then estimating the proportion of the population of the source unit that must be allocated to each target unit based on the area of intersection and a variety of other statistical information programmed into the Great Britain Historical GIS. Figure 6 applies the process to the data on young women in Gloucestershire, interpolating the census data for all three years onto 1911 registration districts.[3]

Geographical standardization may be one of the most important aspects of

Figure 6. Women aged 15 to 24 as a percentage of total population in Gloucestershire using standardized administrative units

By interpolating the data onto target districts (here, 1911 registration districts), spatial patterns can be more accurately compared over time. The technique allows us to compare Bristol in 1881 with Bristol in 1911 and 1931. It also removes the historical artifact of boundary changes, visible in figure 5, that obstructs visual and statistical comparison.

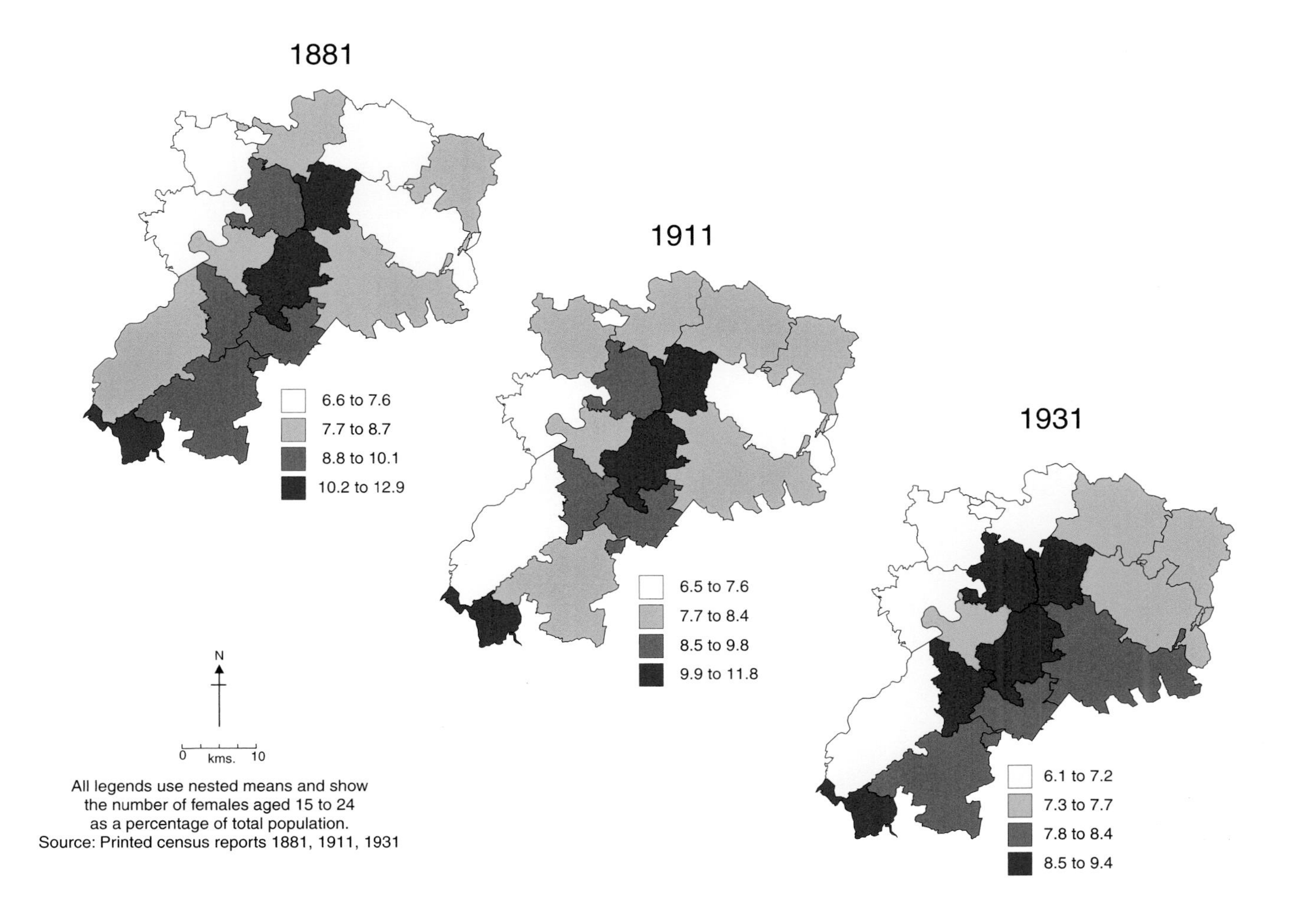

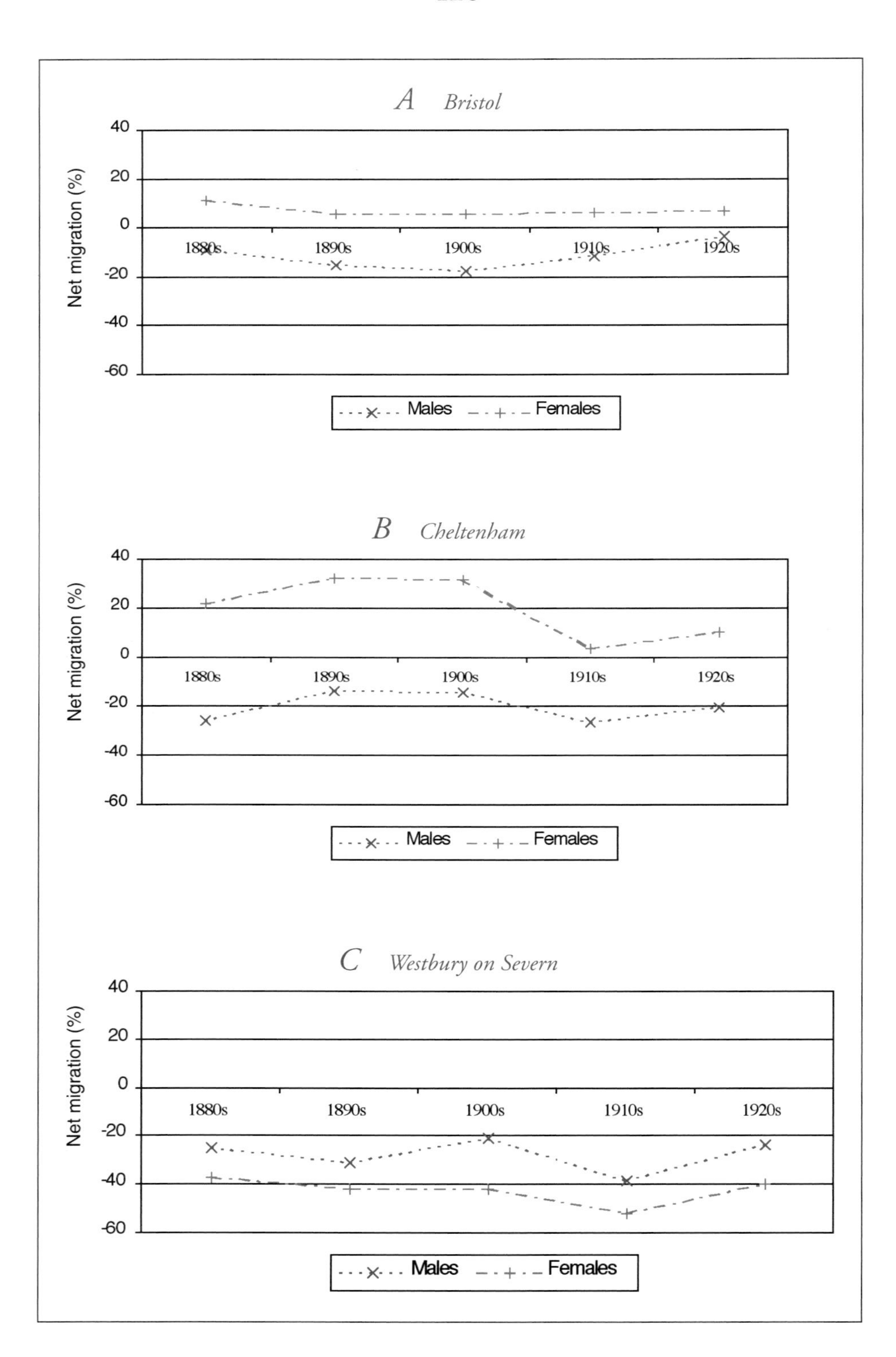

the avenues of research opened up by the project, for it allows us to create long-run time series of spatially referenced data. It also opens up new potential for analysis. Calculating net migration rates offers a good example of this. Net migration is usually calculated by subtracting the number of people in a given place from the number who were there the previous decade, and then subtracting from the residuum the number of deaths over the decade. The number left equals the net migrant population. For example, if Gloucestershire had 30,000 females in 1891 and 28,000 in 1881, and if 1,000 females died in the county between those two dates, the net migration would be (30,000 – 28,000) – 1,000 = 1,000 females. (In this hypothetical example, the result represents net in-migration because the number is positive. A negative number would represent net out-migration.) In the past, this calculation could not be considered accurate because population change caused by boundary changes would appear as net migration. To get around this problem, researchers have resorted to other sources of data that are often available only at county level (e.g., for the whole of Gloucestershire) and that do not distinguish migrants by

Figure 7. Net migration for the cohort aged 5 to 14 in sample target districts, 1881–1931
Net migration is the increase or decrease in population that is not caused by births or deaths. Among young people in Gloucestershire, net migration varied widely in different districts. Bristol, the county's largest city, had modest net male out-migration and female in-migration; Cheltenham, a medium-sized town, had higher rates of male out-migration and female in-migration; while Westbury on Severn, a rural area, had high rates of out-migration for boys and girls.

age or sex. Using the historical GIS allows us to examine geographical change for any and all census categories at any time, and to tease out important differences in the migration patterns of various groups and kinds of places.

Figures 7 and 8 demonstrate the increased level of detail that the GIS can provide. Figure 7 shows time series of net migration rates for males and females who were aged 5 to 14 in 1881. Bristol had low rates of net out-migration for males and low rates of net in-migration for females. Rates of net out-migration were high for both sexes in Westbury on Severn, a rural district, but the female rate was consistently higher than the male. In Cheltenham net female in-migration was very high, presumably reflecting response to the town's growing demand for domestic servants. Figure 8 shows net migration rates for males and females of all ages from 1881 to 1891. Girls had exceptionally high rates of migration into Bristol and Cheltenham and out of Westbury on Severn. Boys and young men also left the rural district at high rates, but they left Bristol at about the same rates. The rate of migration in all districts peaked in youth and declined with age. While historians would expect young people to be most likely to migrate,

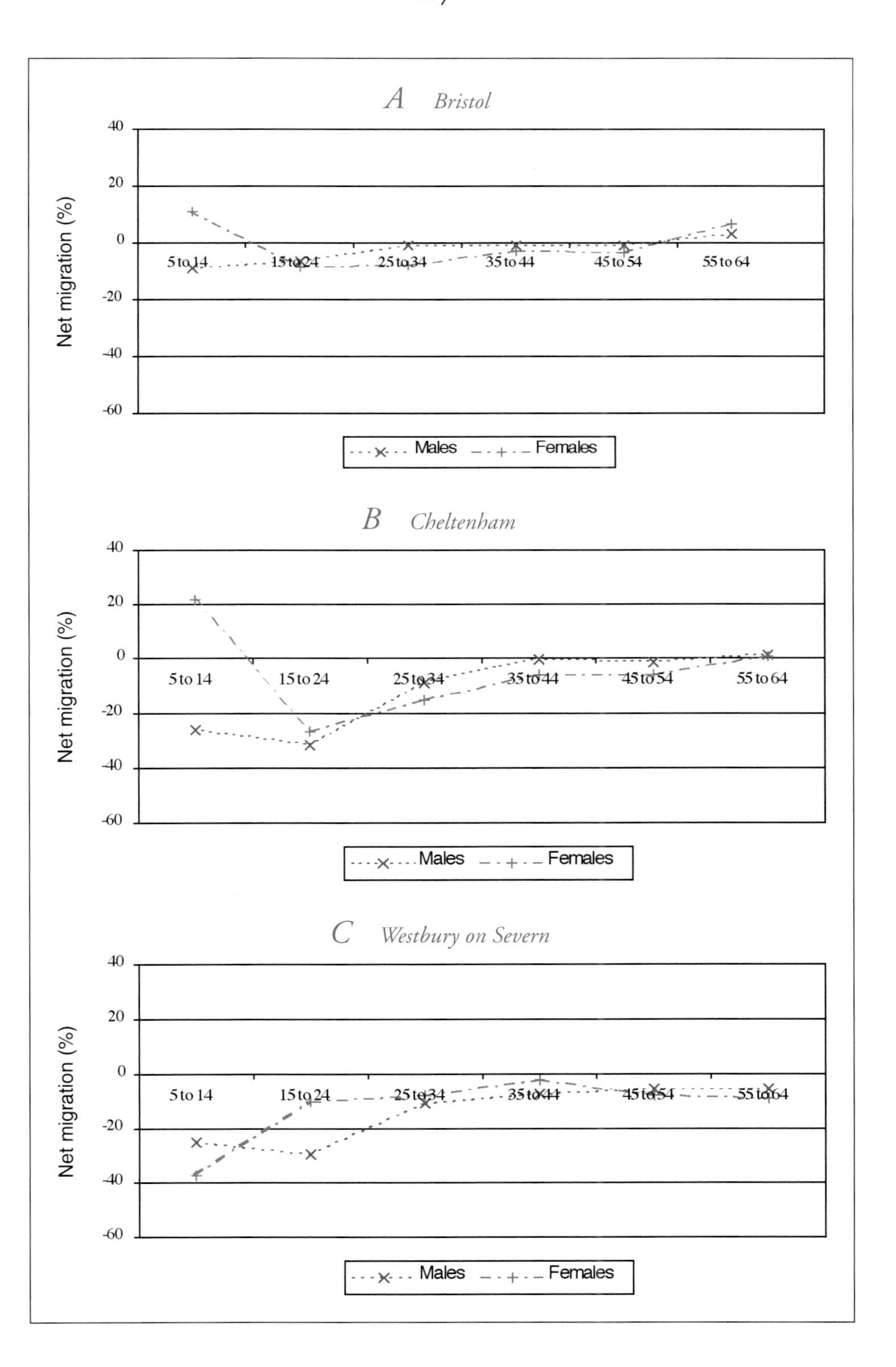

Figure 8. Net migration by age and sex in sample target districts, 1881–1891
This shows that the highest net migration rates were found among the young. Rates for females tend to change more significantly with age than rates for males.

our research is the first to demonstrate the trend with data on the whole population subdivided by age and sex with this degree of spatial detail. Earlier studies have either used the census at a more aggregate level or have had to rely on sample populations drawn from trade union membership lists or genealogical records.

In addition to mapping and spatial analysis, GIS enables historians to visualize data in ways that few but trained cartographers have previously attempted. Most maps displaying historical socioeconomic data have been *choropleth* maps, in which classes of data are displayed by territorial units, such as counties or states. Although choropleth maps are simple to make and easy to understand, they are problematic for mapping population data because large areas with small populations appear disproportionately significant. In British population maps, large rural districts visually dominate smaller but much more densely populated urban districts. One way of redressing this problem is to use *cartograms,* in which the area of each administrative unit is made proportional to its population. Abandoning true shape or replacing it with circles or squares creates an odd-looking map that can be disconcerting upon first glance. Figure 9 shows choropleth and cartogram representations of infant mortality between 1891 and 1901. In the choropleth map, the rates for London are buried beneath crowded boundary lines. The cartogram clarifies urban patterns and reduces the prominence of rural areas. Making cartograms used to be time-consuming, but a GIS can as readily produce a cartogram as a choropleth map, a chart, or any of a number of other modes of representation.

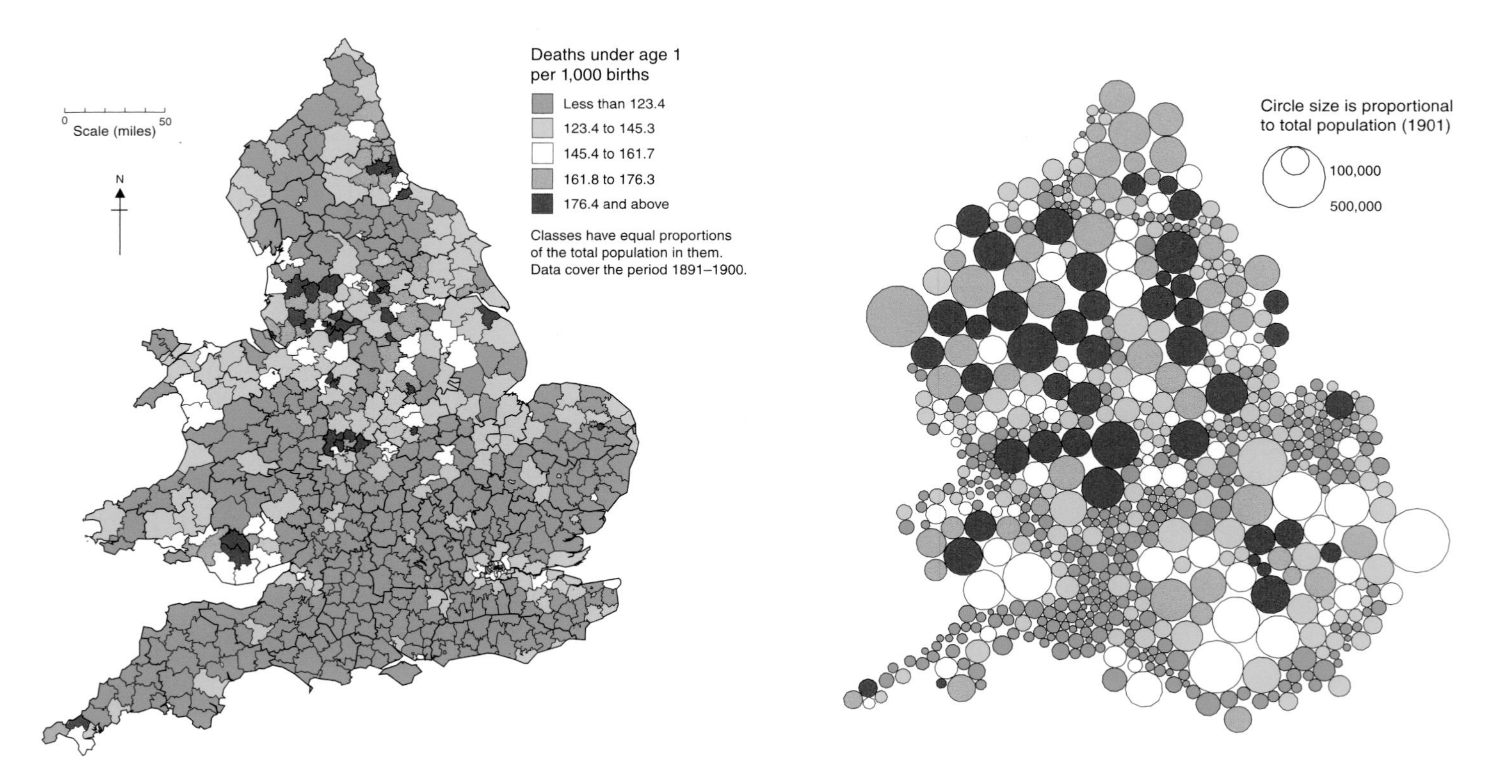

Figure 9. Choropleth and cartogram representations of infant mortality, 1891–1901
The choropleth map on the left overemphasizes areas with low infant mortality rates, most of which are large, sparsely populated rural districts. The cartogram on the right scales each district according to its population size. It shows that far more people were living in districts with high infant mortality than the choropleth suggests, and that infant mortality was chiefly a problem in urban areas at this time.

Researchers have just begun to explore the potential of the Great Britain Historical GIS to unlock the geography of British population history. Combining statistical data with its territorial units creates new analytical opportunities. Being able to visualize every category of data linked to administrative boundaries will raise many new questions and challenge some long-standing assumptions in British demographic history, such as the causes and effects of declining mortality rates from 1851 to 1951. GIS also redefines the role of the map in historical analysis. In the past, a historian would analyze very limited amounts of geographically located data and then send the results to a cartographer for mapping. Using GIS puts mapping at the core of research and data exploration. We will see the past differently as a result.

Acknowledgments

The authors gratefully acknowledge Anne Knowles's extensive contribution to this text. We are also very grateful to Chris Bennett and Vicki Gilham for their dedicated and thorough work on the construction of the Great Britain Historical GIS. The project was primarily funded by the Economic and Social Research Council through grants R000221314, R000221703, R000237506, and R000237757, but with significant contributions from the Leverhulme and Wellcome Trusts and many other funding bodies.

Software and hardware

The Great Britain Historical GIS uses ARC/INFO 7 to store spatial data and Oracle® 7 to store the attribute database. The system is hosted on a Sun™ SPARCstation™ 20 running UNIX. We have also successfully run the system on a Pentium II PC running ARC/INFO 7 for Windows NT with the Oracle database hosted on a remote UNIX server.

Digitizing was done using a CalComp® A0 digitizing table. Maps were plotted on tracing paper using a Hewlett-Packard® DesignJet™ 750C color plotter.

Further reading

Dorling, D. "Area Cartograms: Their Use and Creation." *Concepts and Techniques in Modern Geography* 59 (1996).

Flowerdew, R., and M. Green. "Areal Interpolation and Types of Data." In A. S. Fotheringham and P. A. Rogerson, eds., *Spatial Analysis and GIS,* 121–45. London: Taylor & Francis, 1994.

Garrett, E., A. Reid, K. Schurer, and S. Szreter. *Changing Family Size in England and Wales: Place, Class and Demography 1891–1911.* Cambridge: Cambridge University Press, 2001.

Goodchild, M. F., L. Anselin, and U. Deichmann. "A Framework for the Areal Interpolation of Socio-Economic Data." *Environment and Planning A* 25 (1993): 383–97.

Gregory, I. N. "Longitudinal Analysis of Age and Gender Specific Migration Patterns in England and Wales: A GIS-Based Approach." *Social Science History* 24 (2000): 471–503.

Gregory, I. N., and H. R. Southall. "Spatial Frameworks for Historical Censuses: The Great Britain Historical GIS." In P. K. Hall, R. McCaa, and G. Thorvaldsen, eds., *Handbook of Historical Microdata for Population Research,* 319–33. Minneapolis: Minnesota Population Center, 2000.

Kingdom, J. E. *Local Government and Politics in Britain.* Hemel Hempstead: Philip Allan, 1991.

Langran, G. *Time in Geographic Information Systems.* London: Taylor & Francis, 1992.

Lawton, R. "Population and Society, 1740–1914." In R. A. Dodgshon and R. E. Butler, eds., *An Historical Geography of England and Wales,* 2d ed., 285–321. London: Academic Press, 1990.

Lipman, V. D. *Local Government Areas 1834–1945.* Oxford: Basil Blackwell, 1949.

Oliver, R. R. *Ordnance Survey Maps: A Concise Guide for Historians.* London: Charles Close Society, 1993.

Peuquet, D. J. "It's about Time: A Conceptual Framework for the Representation of Temporal Dynamics in Geographic Information Systems." *Annals of the Association of American Geographers* 84 (1994): 441–61.

Notes

1. In this project we have traced boundary changes back as far as is possible through the use of good-quality maps and nationally produced lists of boundary changes. This has allowed us to trace registration district and Poor Law union boundaries back to 1840 and parish boundaries back to 1876. Researching earlier times would be far more difficult and would take many years of research in local archives using sources whose quality would vary significantly across the country.

2. Strictly speaking, the smallest units for which census data can be mapped are enumeration districts. Reconstructing these boundaries would be a massive undertaking as they have rarely been properly mapped. We have therefore used the parish as our most detailed geography.

3. As areal interpolation involves estimating populations based on areas, population densities, and other information, it inevitably introduces some error to the resulting data. The impact of this error will vary depending on a number of factors, but however slight, it should be considered when interpreting the data. See the further reading for more details of our research on this.

CHAPTER TEN

GIS IN ARCHAEOLOGY

Trevor M. Harris

ARCHAEOLOGY and geography have much in common. Like geographers, archaeologists study places. They collect data in the field and map their findings at many different scales, from a square meter within an archaeological dig to regions covering half a continent. Since the discipline's beginnings in the nineteenth century, when researchers sketched artifacts and landscapes in pencil and watercolor, archaeologists have always relied on visual methods to record and analyze information. During the past decade, they have adopted GIS and GPS to map sites with new precision and to predict where undiscovered sites of archaeological interest might be. The most advanced applications of GIS technology in archaeology use computer simulation to visualize sites in three and four dimensions, giving researchers the ability to peel back the layers of time in the laboratory, with minimal excavation of physical remains. Time and space are intricately connected in the archaeological record. The location, orientation, and depth of objects and sites provide important evidence about their cultural significance and when they were made or were in use. Drawing on examples from my work with collaborators in Great Britain and the United States, the case studies summarized here demonstrate how archaeologists have moved from using GIS to organize and map data to using it for spatial analysis and modeling and for protecting archaeological sites from damage by public works projects. This chapter reports on the use of GIS for archaeological mapping and 2.5-D display, viewshed analysis, the exploration of relative distance and friction surfaces, true 3-D multidimensional models, archaeological site prediction, and virtual GIS and virtual world applications.

Archaeologists first used GIS to collate data from paper maps and other sources digitally and to reorganize it within a database structure. Early archaeological GIS use focused on the reproduction of printed archaeological maps in that they gathered large quantities of spatial and temporal data into a few layers, or coverages. Once data is integrated within a GIS it is much easier to disaggregate, recombine, and display selected features or classes of information than it is to extract them from a printed map. Gary Lock and I used these GIS capabilities to reexamine archaeological information about the Danebury hillfort and its surrounding landscape.[1] Danebury is one of the most intensively studied prehistoric hillfort sites in Western Europe. Its archaeological record extends from the Paleolithic to the Romano-British period. We focused primarily on the period following the Neolithic, some fourteen thousand years B.P. (Before Present). Much of the information for Danebury came from excavation reports, aerial photographs, and site records that had been compiled in maps that inventoried four broad time periods of the site's archaeological record. We digitized these maps and combined them with digital coverages of present-day hydrology, soils, and topography.

With the resulting GIS database we were able to dissect the landscape surrounding Danebury digitally. For example, we asked the system to show the location of woodlands, buildings, and other modern features on the modern topographical maps that might have covered or disturbed the archaeological record *(figure 1)*. By adding elevation data and displaying archaeological sites as three-dimensional images,[2] we could more readily see how the location of ancient features related to topography and how the human landscape changed over time *(figure 2)*. In contrast to the communal burial traditions of the Neolithic period, for example, Bronze Age people buried their dead in individual round barrows that represented both personal wealth and the increasing importance of social status. The proliferation of burial sites in this period, their gradual clustering into cemeteries, and the clearing of woodland to make way for barrows and cemeteries, demonstrates a much more organized use of the landscape. In the Late Bronze Age, landscape organization shifted from symbolic and ceremonial purposes to agriculture as farmers built settlements, fields, and ditches that were also used as routeways and boundaries. Hillforts, which appeared in the Iron Age, established nodes of military, economic, and political control. Archaeologists have known for some time that hillforts around Danebury, as elsewhere, were built gradually over several hundred years. Previous maps, however, collapsed the period of construction into a single image. With GIS we were able to see that there was indeed a sequencing of hillfort construction and use, which helped us greatly in understanding not only the central place hierarchy of these forts but also the varying changes occurring in the surrounding archaeological landscape.

Viewshed analysis is a GIS technique that enables one to simulate the effect of standing at any point in the landscape and seeing what one would actually see from that location.[3] We used viewshed

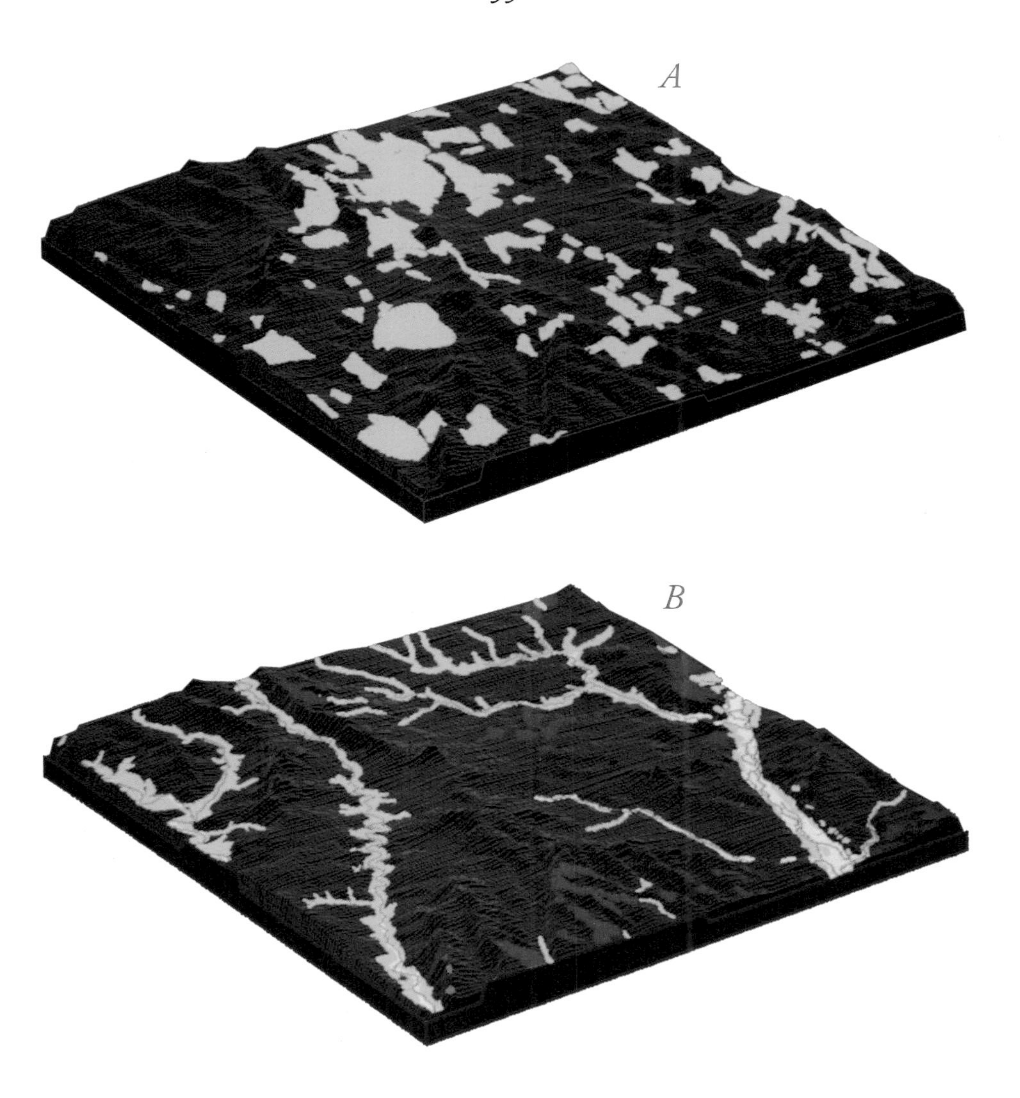

Figure 1. Select views of Danebury through the lens of GIS
By highlighting selected features within the GIS database, one can single out particular features in the landscape for examination. Map A highlights areas where the archaeological record may have been disturbed, destroyed, or covered by modern construction or woodland. Map B identifies the location of streams and fertile soils, where one might expect to find archaeological remains associated with ancient settlements.

analysis to examine the significance of long barrows, one of the few kinds of evidence that survives from the Neolithic period in the Danebury area. Long barrows were known to be funerary monuments—larger versions of round barrows—as well as being territorial markers dominating a specific landscape. This interpretation was supported by the fact that some barrows contain no human remains, and the visual dominance of a long barrow over a particular area was interpreted as representing the demarcation of a prehistoric group's social and political territory. Our viewshed analysis confirmed that barrows could indeed have served as territorial markers, for some barrows came into view as one crossed a

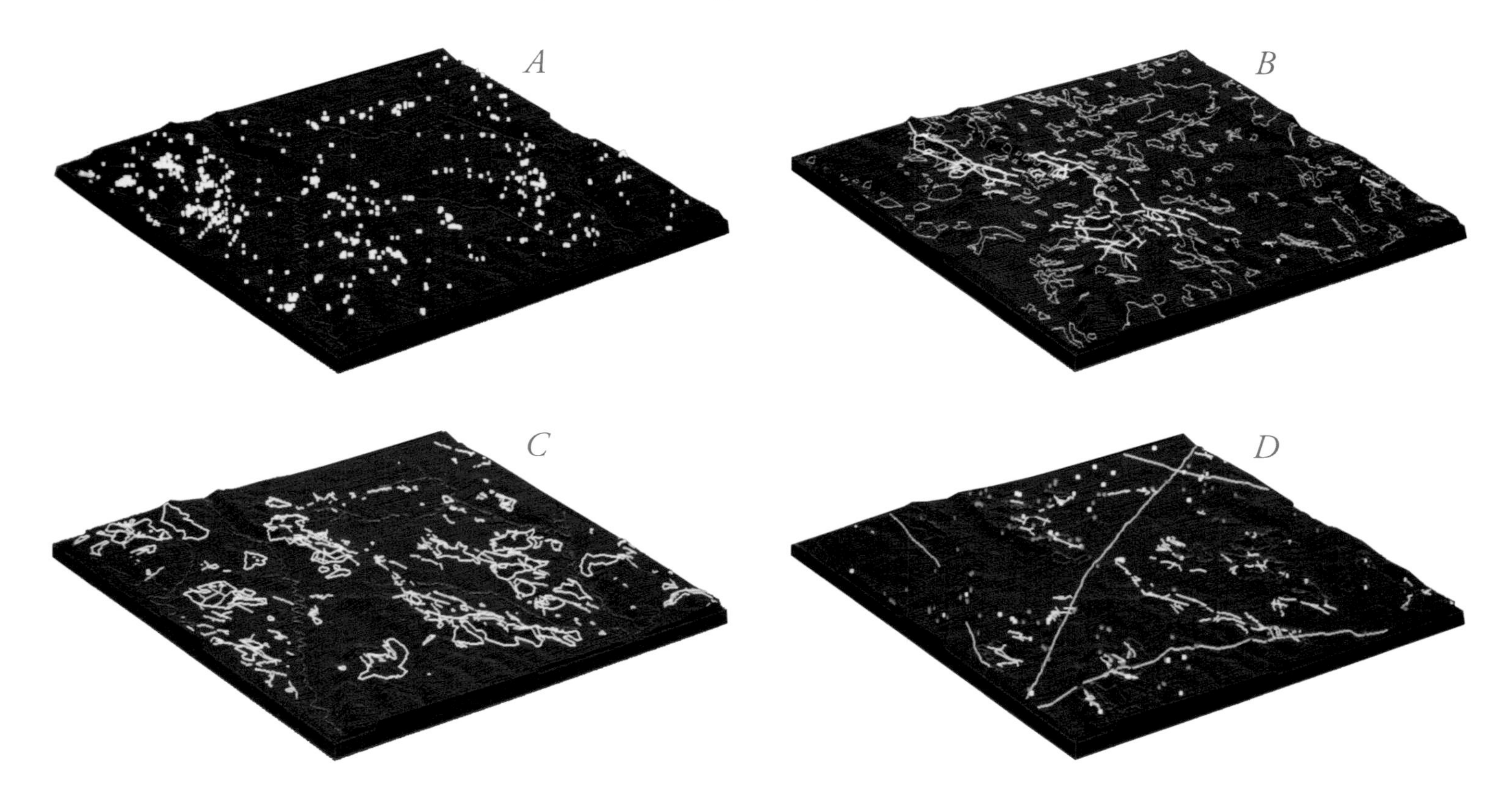

Figure 2. Four eras of the Danebury landscape

The most distinctive landscape features in the Bronze Age (map A) were round barrows. In contrast to the communal burial traditions of the Neolithic period, Bronze Age people buried their dead in individual round mounds that represented both personal wealth and social status. In the late Bronze Age there is evidence of a change in landscape use from funerary and ceremonial purposes to settlements, small fields, and linear ditches (map B). As population grew and more intensive farming depleted soils, burial sites were no longer off limits for agriculture and Iron Age people transformed the landscape with farmsteads, field systems, ditches, and hillforts (map C). Under the Romans the landscape underwent rapid reorganization as new roads were constructed heedless of the earlier centers of military, economic, and political control (map D).

ridgeline *(figure 3)*. Travelers walking to the crest of those hills would have known, the theory goes, that they had crossed a boundary when the next barrow came into view. But our analysis also showed that in the Danebury region these barrows were invisible from each other. In each and every case, the viewshed of one long barrow did not overlap with the viewshed of any other long barrow. The barrow builders used topography in remarkably subtle and intricate ways. Their deliberate concealment of barrows from the view of neighboring groups suggests that territories were more complex than we realized and that we may not yet fully understand the locational, symbolic, or functional significance of the barrows.

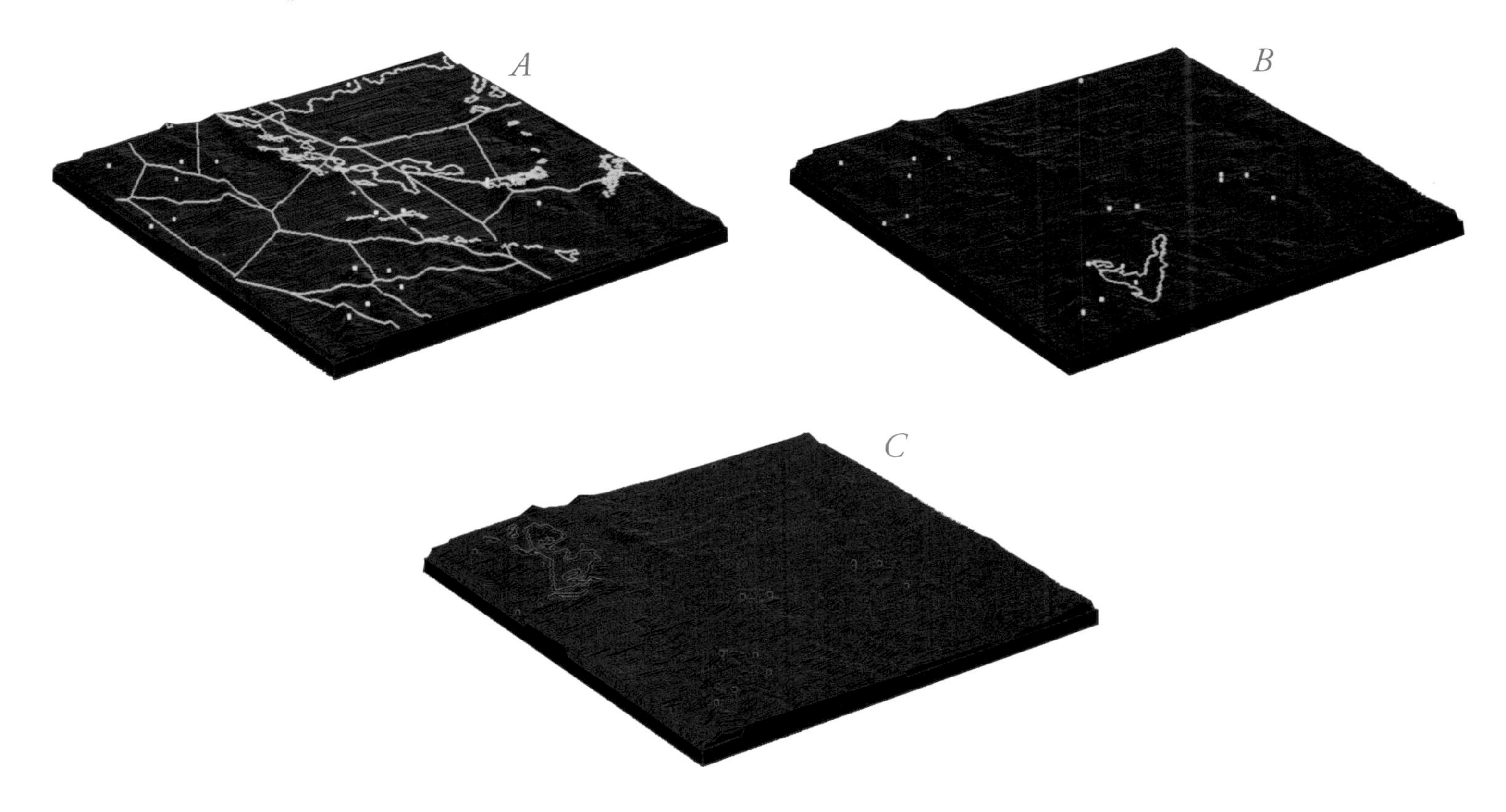

Figure 3. Viewshed analysis of Danebury long barrows

Thiessen polygons place boundaries at equal distances between adjacent long barrows and are valuable in helping to demarcate "market" or territorial boundaries (map A). In this instance, however, the polygons provide little evidence of a regular territorial division of the prehistoric Danebury landscape. The viewsheds of a few selected long barrows (maps B and C), outlined in bright green, show that the areas from which the monuments were visible varied in size and shape. Some viewsheds followed ridgelines, but some did not. Remarkably for the Danebury region, no viewshed from a long barrow overlapped with the viewshed of any other long barrow.

We also used GIS to study the siting of Iron Age hillforts. The conventional interpretation is that hillforts were located on ridges and hilltops to command the broadest view of the surrounding country and its inhabitants, and to provide a safe retreat for the people who farmed in their shadow. A viewshed analysis showed that almost all the farmsteads surrounding Danebury could indeed be seen from the hillfort. But Iron Age inhabitants traveling on foot would traverse the area's hills and valleys with varying degrees of difficulty. To estimate the fort's accessibility we created a *friction surface* that calculated the time and difficulty of travel over a given area by modeling the impediments posed by slope, distance from the hillfort, water bodies and streams, and soil type. The resulting image *(figure 4)* shows that accessibility did not vary uniformly with distance. It would have taken longer for some people living near the fort to reach it than others living much farther away. Seeing the landscape in this way helped us understand the distribution and placement of farmsteads around Danebury as well as other hillforts whose presence has previously been poorly explained. The distinction between absolute distance (the distance traversed on a map "as the crow flies") and relative distance (the time or effort that it took to traverse a landscape) provides a valuable insight into the actions of prehistoric peoples. An analysis of the relative proximity of sites and their interaction in the Danebury region takes on an additional but more realistic complexity. Such a concept, for example, helped to explain the location of another hillfort to the east of Danebury that, given the size and importance of Danebury, is surprising to see so close by. The friction surface indicates that the intervening marshy valley was difficult and time-consuming to cross and required that another hillfort be situated in this area to control and support the local inhabitants.

Most GIS applications and conventional mapping methods impose a two-dimensional abstraction of reality, seriously constraining one's ability to analyze and represent time and depth. Archaeologists have usually dealt with this problem by stacking vertical layers of two-dimensional distribution maps, as in our Danebury example. Most historical GIS has incorporated time simply as one of many fields in the attribute database, where it can be used to examine distribution patterns based on a specified date or period. True three-dimensional topology would enable depth and time to be treated as independent axes. Thus where a buffer in a two-dimensional GIS might be a circle, in a three-dimensional system the

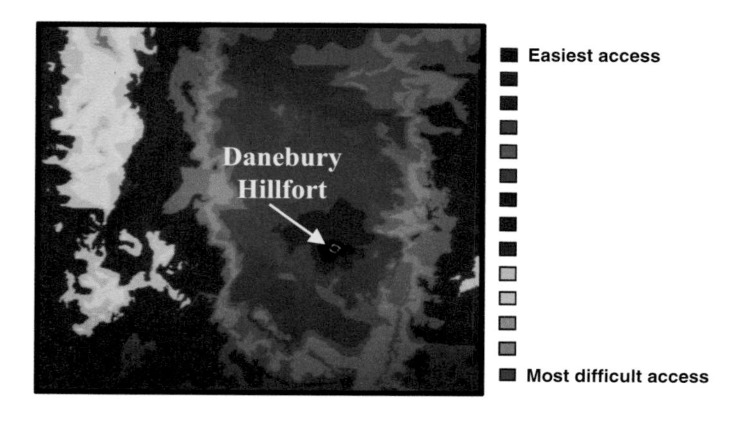

Figure 4. Accessibility of Danebury hillfort

Based on the combined influence of slope, distance, soils, and the presence of water, it can be seen that Danebury was more accessible from some surrounding areas than from others. Areas shaded blue and black were easier to traverse than those shaded magenta or red. Areas shaded yellow to green were the most difficult to traverse. Farmsteads were primarily clustered in the more readily accessible zones.

buffer could be a spheroid, cone, or tube. Systems with this capability have recently become available, borrowing 3-D modeling techniques developed for the commercial exploration of petroleum and gas reserves, hydrogeology, and contaminant plume analysis.

Three-dimensional models have rarely been used in archaeology. Of those that have, the most common extends the four-sided, four-node pixels of a 2-D raster GIS into six-sided, eight-node cubes or voxels.[4] With three quantified dimensions, archaeologists can create a three-dimensional representation of an archaeological site that can be "sliced" through in any direction and depth to expose interior features. Alternatively, isosurfaces can be created that display a class of features within a specified period of time or selected range of depths. While it is clearly an oversimplification to equate time with depth in archaeology, there is a linkage between the two in that, in general, the deeper an artifact or find is in an archaeological series, the older it can be assumed to be. Three-dimensional GIS was applied to a survey of a suspected Romano-British settlement in England. To determine whether the site was in fact previously inhabited, bore holes were driven and extracted soil samples across the site were analyzed for phosphate concentrations. Since fires, food refuse, and human waste leave concentrations of phosphates in the soil, the measure is a good indicator of human activity. Figure 5 shows the results of our analysis. The 3-D images represent layers of soil, or soil horizons, from the present day through to the Romano-British period. In this study, 3-D GIS helped us see the depth and location of phosphate concentrations. This enabled researchers to not only verify site occupancy but to understand the varied uses of the site over the past two thousand years.

Government agencies in many countries rely on archaeologists to identify sites that may be threatened by urban–industrial growth, new roads, dams, and other construction projects. Because the archaeological record is invariably incomplete, it is often necessary to use predictive modeling to determine whether a proposed development is likely to damage valuable national resources. In 1993, the American Electric Power Company proposed erecting a 765-kV high-power transmission line from Wyoming County in West Virginia to Roanoke County in Virginia, a distance of some 120 miles. As part of the project's environmental impact assessment, researchers at West Virginia University modeled

Figure 5. Using 3-D imaging to pinpoint human habitation
Phosphate levels in soil are a good indication of former human habitation. At a suspected Romano-British site, strong phosphate concentrations were found (map A, light blue, green, and yellow) in the 2.5-meter soil horizon. (The vertical scale is in centimeters.) Slicing the 3-D image at a variety of vertical and horizontal cross sections indicates the location of these concentrations. Slicing the image by a phosphate threshold value (map B) to create an iso-surface shows that the deposits are not uniform across the area and points to more intensive occupation of certain parts of the site.

A

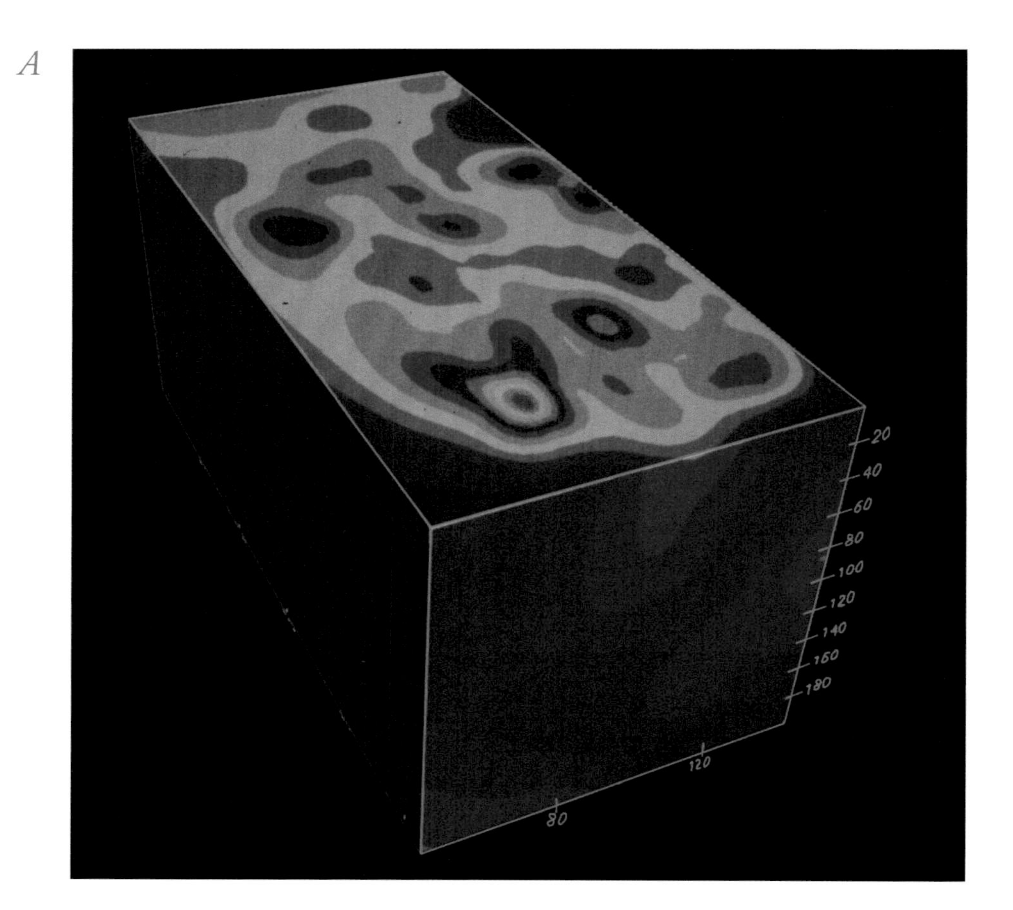

B

the power-line corridor to see how it would affect archaeological sites. We analyzed the data for known archaeological sites within the study zone, including elevation, slope, distance from water, aspect, and soil type. By searching the entire corridor for locations with similar characteristics, the model predicted the probability of finding various kinds of archaeological sites in unexplored locations *(figure 6)*. Based on known archaeological remains, further finds were considered highly probable in locations within 230 meters of a water body or stream, on a slope of less than 18 degrees, at an elevation of between 400 and 600 meters, on well-drained soils. Although creating the GIS took considerable effort, it greatly narrowed the search area and targeted sites that the two states' cultural resource management agencies could investigate without severely straining their limited budgets. It also created a body of knowledge that can benefit future projects.

Environmental factors can account for only so much of human history or prehistory. But because it can be very difficult to incorporate qualitative cultural data in predictive models, many GIS applications in archaeology rely heavily on available digital environmental data, raising concerns about a return to environmental determinism. Incorporating cultural understanding into a GIS is not straightforward, but it can be done. A group's tradition of worshiping the dead and of settling near burial sites, for instance, could be factored into a buffer analysis to predict the probability of finding settlements within certain distances of a known burial site. The model could also include distance to streams, elevation, and other physical factors known to influence settlement patterns. One would expect a model combining environmental and cultural factors to yield better predictions than either taken in isolation.

In ongoing research on the Grave Creek Mound in Moundsville, West Virginia, Jesse Rouse and I are seeking to embed qualitative information within a GIS. Grave Creek Mound, which dates to 2500 B.P., is the largest extant burial mound of the Adena people. The aim of our project was to integrate the geospatial data we gathered for the site with cultural resource information found in print and manuscript sources, and to combine them with GIS functionality and mapping for distribution to a global audience. The Internet GIS contains maps, photographs, moving images, text, narrative, oral histories, and dynamic imagery developed with virtual GIS.

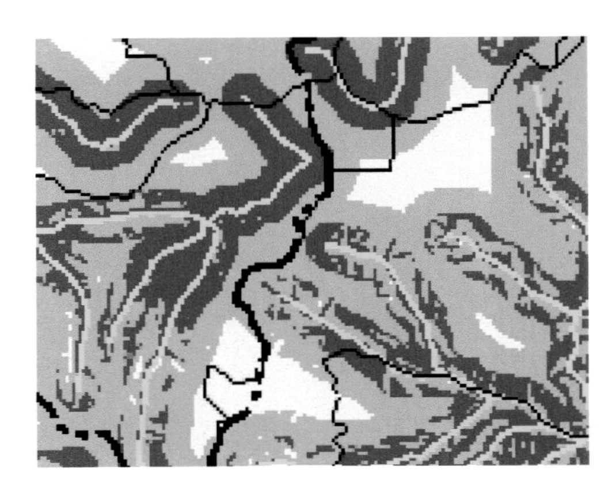

Figure 6. Predicting archaeological sites

Sometimes major public works provide incentive and funds for new archaeological research. The environmental impact assessment of the American Electric Power Company's plan for a new power corridor required detailed modeling of archaeological sites along the corridor's 120-mile path. Based on the modeling of environmental factors that influenced the location of known archaeological sites, the red areas on this detail of our map show where the model predicts the greatest likelihood of finding archaeological remains.

Virtual GIS combines analytical GIS with virtual reality. Like virtual reality, virtual GIS simulates the world in three physical dimensions and in time. Figure 7 shows an image from a virtual GIS simulation of a low-level flight down the Ohio River Valley toward Grave Creek Mound.[5] In a simulation based on vegetation data from published pollen analysis studies, we re-created the landscape immediately surrounding Grave Creek Mound as a virtual world as it might have looked when the Adena people built the mound *(figure 8)*. This stretch of the Ohio has been heavily industrialized and urbanized for almost two centuries. Removing these modern developments and repopulating the landscape with vegetation as it might have looked some twenty-five hundred years ago requires processing considerable data, but the effort helps students and scholars better perceive the physical and historical context of the Adena mound-building project. Using virtual reality tools, one can change the angle of sunlight striking the mound through the course of the day, bring trees into leaf in a simulated spring, and strip away the leaves for a winter view. These are not fancy cinematographic effects but techniques that give substance

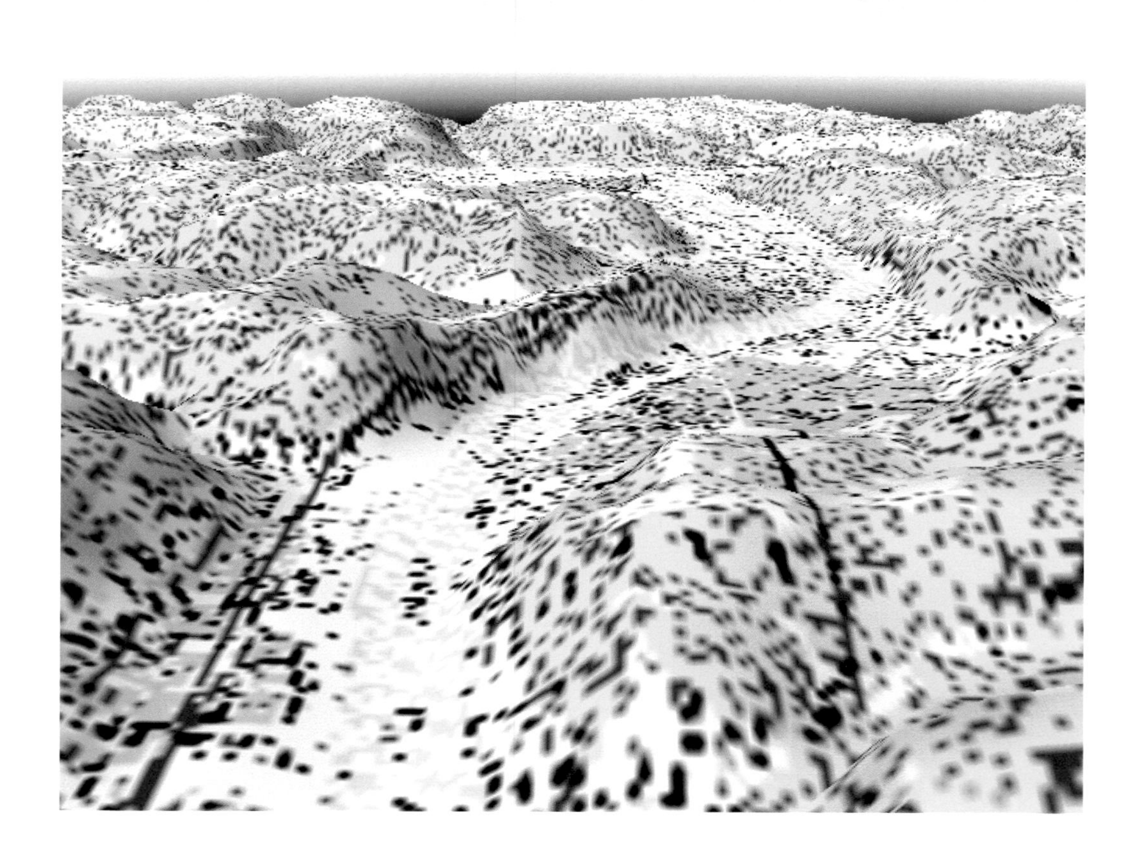

Figure 7. A virtual flight past an Adena burial mound
Combining virtual reality with GIS, as flight simulators do, enables one to move dynamically through three-dimensional environments. To create the terrain model we used the area's 1:24,000 U.S. Geological Survey topographic map, digital aerial photography, and satellite imagery. Towns are shaded pink, the Ohio River is light blue, and the purple lines are roads and railway lines. The mound lies in the bend of the floodplain in the center background of the image.

Figure 8. Simulating the Adena landscape circa 2500 b.p.
The final step in simulation is to strip away modern landscape features and reinstate historically correct vegetation. These frames from a simulated flight skim just above the Ohio River where it bends southwest around present-day Moundsville. The mound is located to the left of the image.

and form to landscape features whose physical presence and effect on prehistoric communities may otherwise be difficult, if not impossible, to reconstruct. The virtual simulation of Grave Creek Mound shows, among other things, that the heavy vegetation of the area's deciduous trees would have limited what was visible from the mound, something one simply would not be able to see in a simpler model.

In moving from static, two-dimensional maps to dynamic three- and four-dimensional visualizations of prehistoric sites, archaeologists have greatly expanded their ability to investigate and analyze. While these methods may all be poor substitutes for experiencing the world as prehistoric peoples once did, they are a big step forward from what was possible just a decade ago.

Further reading

Allen, K. M. S, S. W. Green, and E. B. W. Zubrow, eds. *Interpreting Space: GIS and Archaeology.* London: Taylor & Francis, 1990.

Fisher, P., and D. Unwin. *Virtual Reality in Geography.* London: Taylor & Francis, 2002.

Lock. G., ed. *Beyond the Map: Archaeology and Spatial Technologies.* Amsterdam: IOS Press, 2000.

Raper, J. *Multidimensional Geographic Information Science.* London: Taylor & Francis, 2001.

Lock, G. R., and Z. Stancic, eds. *Archaeology and Geographic Information Systems: A European Perspective.* London: Taylor & Francis, 1995.

Wescott, K. L., and R. J. Brandon, eds. *Practical Applications of GIS for Archaeologists: A Predictive Modeling Toolkit.* London: Taylor & Francis, 2000.

Wheatley, D., and M. Gillings. *Spatial Technology and Archaeology.* London: Taylor & Francis, 2002.

Online resource
Grave Creek Mound Web site: ark.wvu.edu/grave_creek/default.html

Notes

1. Gary R. Lock and Trevor M. Harris, "Danebury Revisited: An English Iron Age Hillfort in a Digital Landscape," in M. Aldenderfer and H. D. G. Maschner, eds., *Anthropology, Space, and Geographic Information Systems,* 214–40 (New York: Oxford University Press, 1996).
2. In the Danebury GIS displays, we used 2.5-D draped images that simulate the three-dimensionality of the terrain surface as visualizations but not as true 3-D GIS representations containing three independent axes (x, y, z). See Jonathan Raper, *Multidimensional Geographic Information Science* (London: Taylor & Francis, 2000).
3. Viewshed analysis employs the topological relationships embedded in GIS, the data within a digital elevation model, and line-of-sight analysis to determine what parts of the landscape can be seen from any given point in the terrain.
4. Alternate 3-D data models are based on surface piecewise patches welded by low-order parametric polynomials or NURBS (non-uniform rational b-splines) and models derived from triangulated tessellation models in 2-D GIS, in which solid triangulated tessellations provide a solid 3-D geometry.
5. See ark.wvu.edu/grave_creek/default.html for online access to this and other images from the virtual GIS of Grave Creek Mound.

CHAPTER ELEVEN

MAPPING THE ANCIENT WORLD

Tom Elliott and Richard Talbert

SUPPOSE that a student preparing a paper on the political and military activities of King Philip of Macedon comes across the following passage in a speech of Demosthenes, the famous Athenian orator:

> *Apparently those who inhabited Amphipolis, before Philip took it, were holding Athenian territory; but when he has taken it, it is no longer our territory, but his own, that he holds; and in the same way at Olynthus and Apollonia and Pallene he is in possession of his own property, not that of others.*[1]

Where would the student turn to find the location of these places and learn about their strategic importance in the fourth century B.C.? Upon consulting a gazetteer of the ancient world, the student might be surprised to learn that there was more than one place named Apollonia. Which one is Demosthenes talking about? Why does he bring it up at all?

Geographical questions would similarly confront a cultural historian studying the Roman conquest of the Po river valley in northeast Italy. A researcher who wanted to analyze how land was redistributed under the Romans and how that process affected the culture of the valley would want to conduct complex spatial analyses to examine temporal and spatial trends. But before any GIS work could begin, it would be necessary to consult literary and documentary sources, aerial photographs, and archaeological surveys. Deciding how to classify the ways the landscape was reorganized, identifying remnants of these patterns in the landscape, and dating them would require data that lies scattered in scores of archaeological investigations conducted over many decades. All these tasks would be complicated by the scarcity of reference maps showing the location of ancient settlements in the region.

For most of the twentieth century, students of antiquity faced serious difficulties if their research interests were geographical. Although nineteenth-century scholars of Greek and Roman history used cartography, the increasing specialization and diversification of academic disciplines distanced ancient history from its spatial context. The one important exception is the field of archaeology, which developed as a discipline almost entirely since the publication of the last comprehensive atlas of the Greek and Roman world in 1874.[2]

Although various scholarly projects since 1900 have mapped particular aspects of ancient society, none collected and distilled enough of the century's considerable published findings to provide a comprehensive source of spatial data on Greek and Roman antiquity. By 1980, the need for such a geographical–historical resource had become so acute that a report on research tools for the classics, published by the American Philological Association (APA), cited cartography as "an area . . . where the state of our tools is utterly disastrous."[3] It was from this public recognition of the problem that the APA's Classical Atlas Project was born. In 1988 the project was headquartered at the University of North Carolina in Chapel Hill under the direction of Richard Talbert.

In fall 2000, the project published the first comprehensive atlas of classical antiquity since William Smith's atlas of 1874 *(figure 1)*. The *Barrington Atlas of the Greek and Roman World,* issued by Princeton University Press, provides 102 full-color maps that show the physical and cultural world of the Greeks and Romans, from the British Isles to the Indian subcontinent and deep into North Africa. The atlas documents more than twenty-five thousand named places and features spanning approximately 1000 B.C. to A.D. 640, together with many more unnamed ancient remains. This human geography is plotted on a landscape from which the effects of modern activity

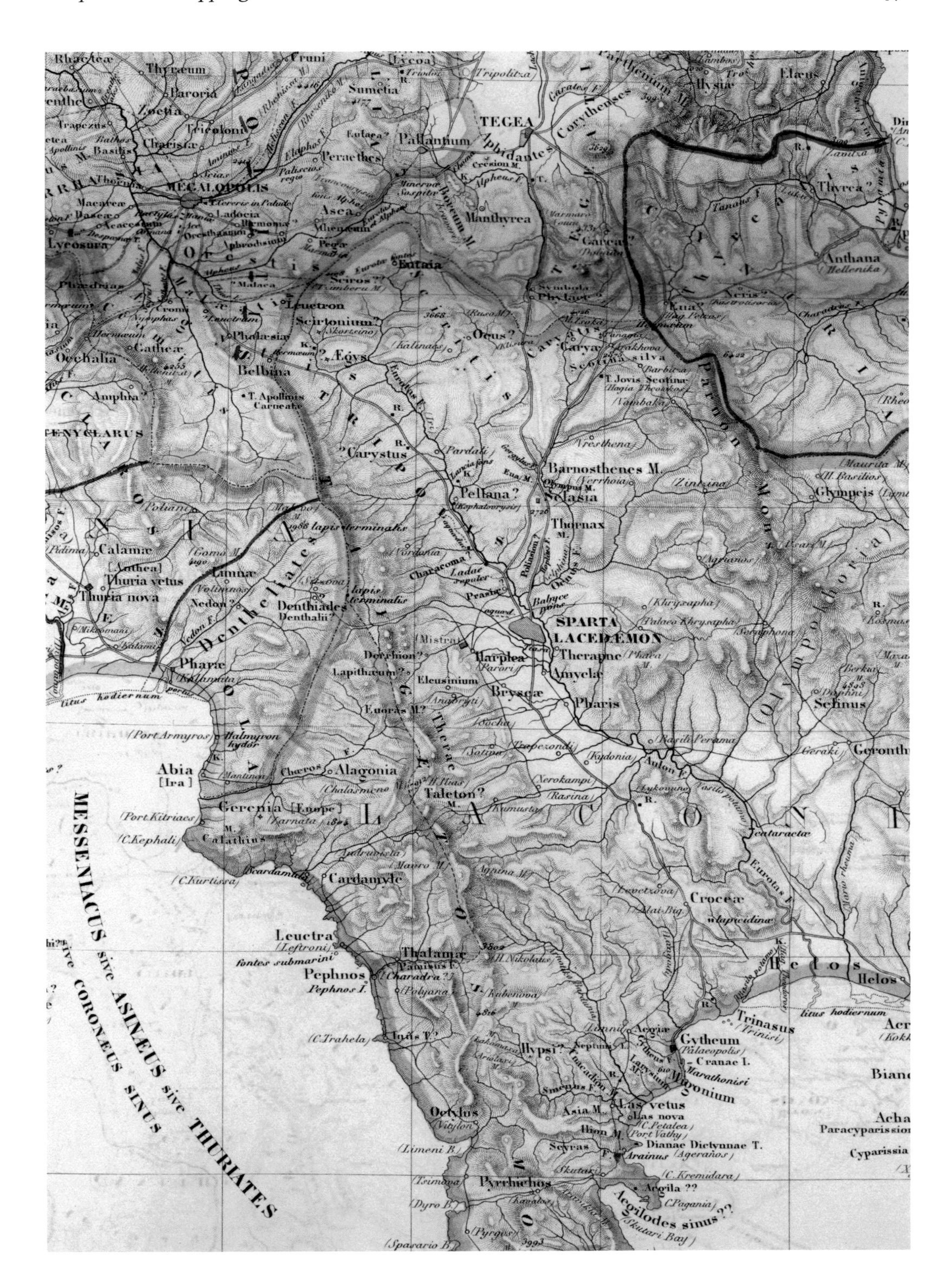

Figure 1. The Peloponnesus in Smith's Atlas of Ancient Geography

Each map in William Smith's 1874 atlas was engraved onto bronze plates, printed, and then colored by hand. Its rendering of shaded relief, rivers, coastlines, site locations, and place-names represented the state of the art in Victorian cartography and classical studies. But the tremendous cost of the volume put it beyond the reach of most students and scholars, and the cost of making corrections or updating the volume were so prohibitive that a second edition was never published.

Figure 2. Locator map from the Barrington Atlas
Colored outlines indicate the coverage and scale of each map in the Barrington Atlas. Modern coastlines and political boundaries, used here to help orient users to the location of individual maps, do not appear elsewhere in the atlas.

and climatic change, such as reservoirs and coastline shifts, have been removed. Twelve years in the making, at a cost of $4.5 million, the atlas employed the services of approximately two hundred compilers, editors, researchers, reviewers, and cartographers, including scholars from universities, museums, and research institutes around the world. Cartographic services were provided by MapQuest.com.

The atlas opens with six maps at a scale of 1:5,000,000 to orient readers to the regions that comprised the Greek and Roman world *(figure 2)*. Shaded relief and *hypsometric* color tinting (which marks changes in elevation) give a broad sense of the terrain of classical civilization *(figure 3)*.

The next ninety-three maps depict the ancient landscape in more detail, showing roads, fortifications, aqueducts, bridges, and other man-made features. For the periphery of the Greek and Roman world, the scale of 1:1,000,000 is employed *(figure 4)*. A more generous scale of 1:500,000 is used for the Mediterranean core, and the finest scale—1:150,000—for the environs of Rome, Athens, and Constantinople, the greatest cities of classical antiquity. Finally, three overview maps depict the provincial boundaries of the Roman empire at different periods.

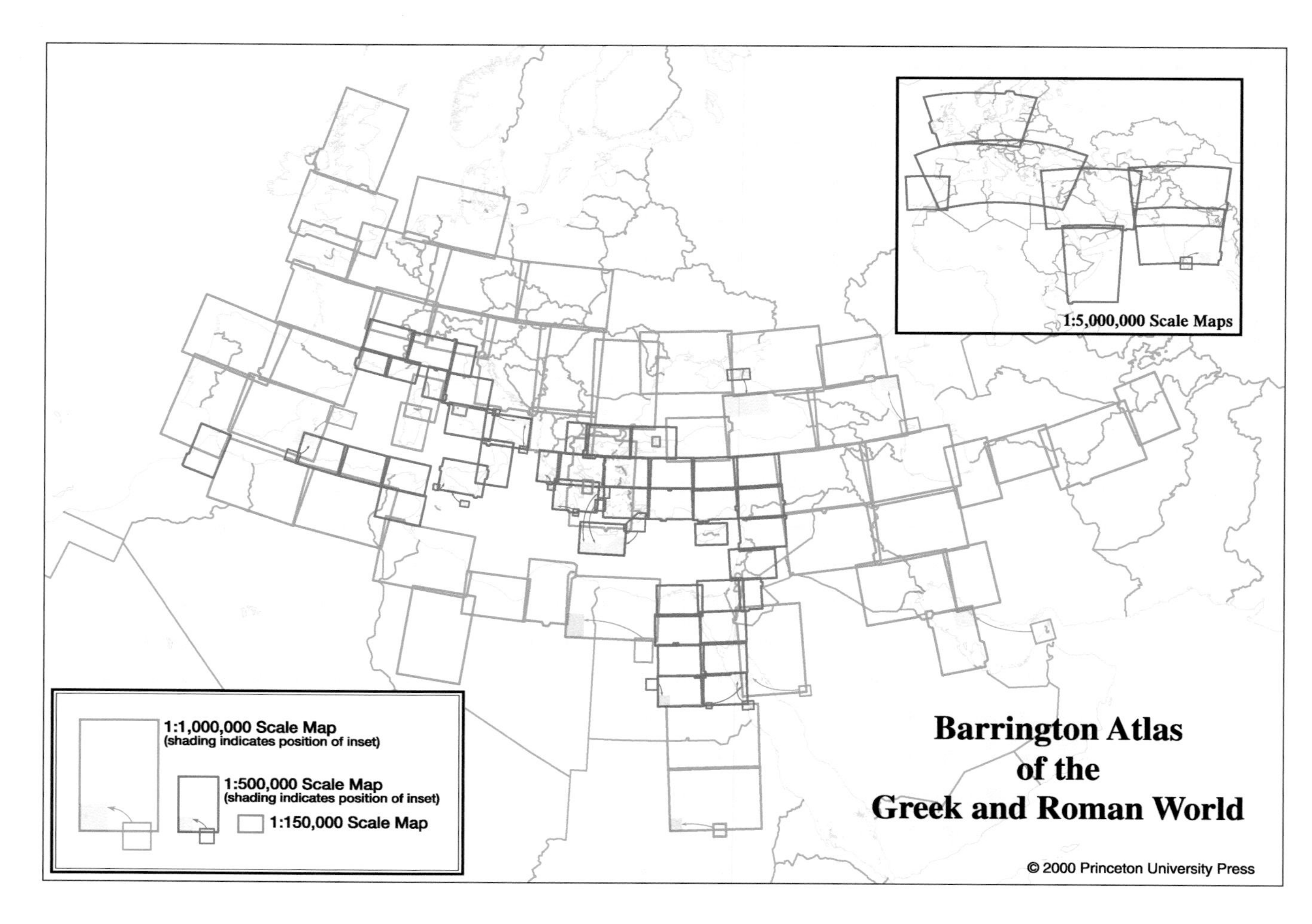

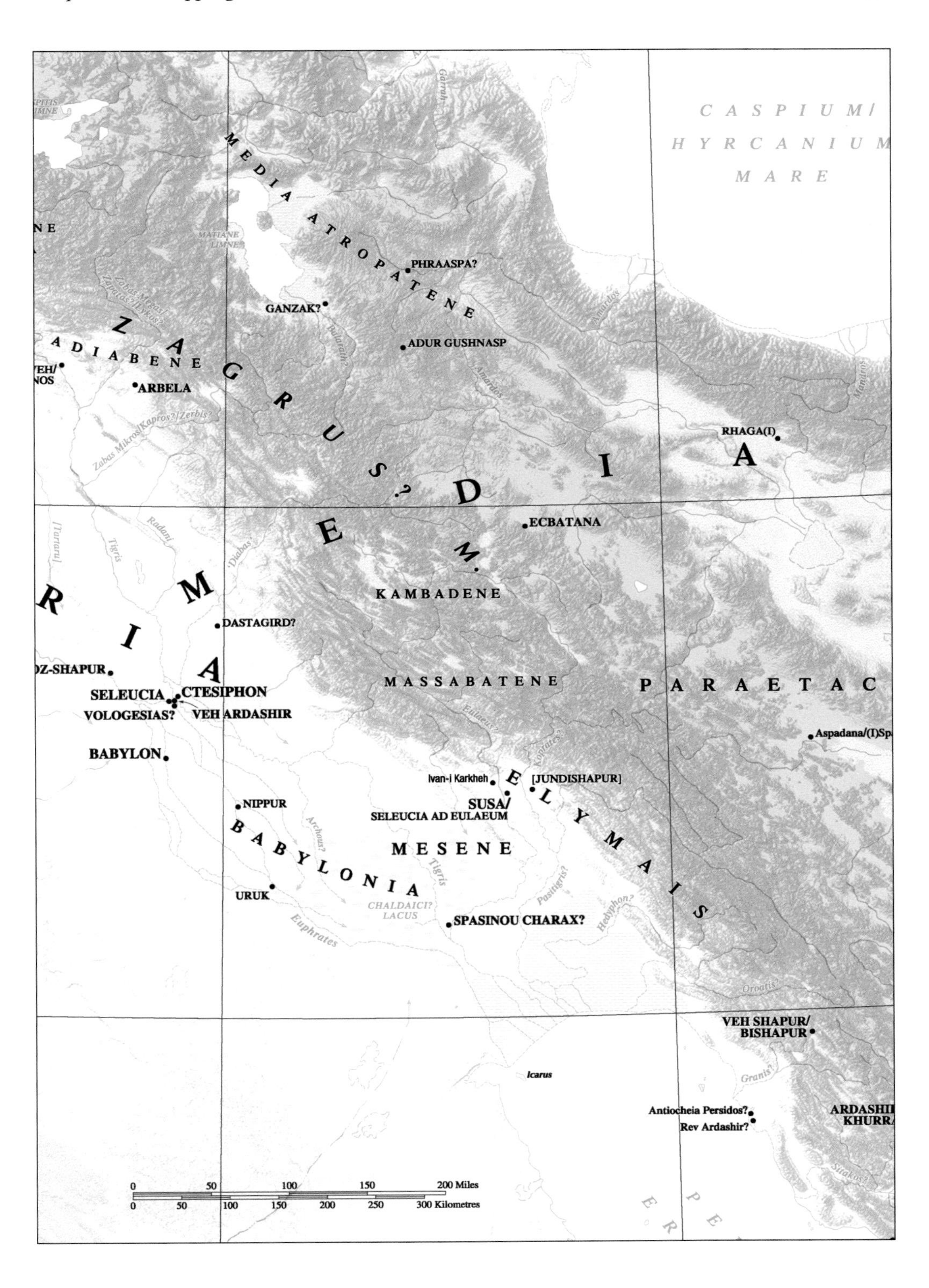

Figure 3. Landscape representation in the Barrington Atlas
This detail of "Asia Occidentalis" shows the shaded relief, elevation tints, and place-names typical of the six Barrington Atlas maps at a scale of 1:5,000,000. This excerpt is centered on ancient Babylonia (modern Iraq, Iran, Saudi Arabia, and Kuwait). It shows portions of the Tigris and Euphrates rivers, the Caspian Sea (upper right), the Persian Gulf (lower right), and the Zagros Mountains. Many of the prominent sites from earlier cultures are not shown because there is no evidence that they were occupied or otherwise important during Greek or Roman times.

Figure 4. The Roman presence along the Danube

This map shows the delta and lower reaches of the Danube River (ancient Danuvius/Istros/Hister) where it empties into the Black Sea. Like other Barrington Atlas maps at the 1:1,000,000 scale, this one provides complete coverage of sites, features, and names in peripheral areas of the Greek and Roman world. The lack of towns and roads as one moves north across the Danube indicates cultural differences in antiquity and the imbalance of scholarly research on this region. The Danube formed a barrier to Roman expansion and a defensive shield for Roman settlements. Most of the fortified sites (black squares) along the road paralleling the river were founded in the Roman period.

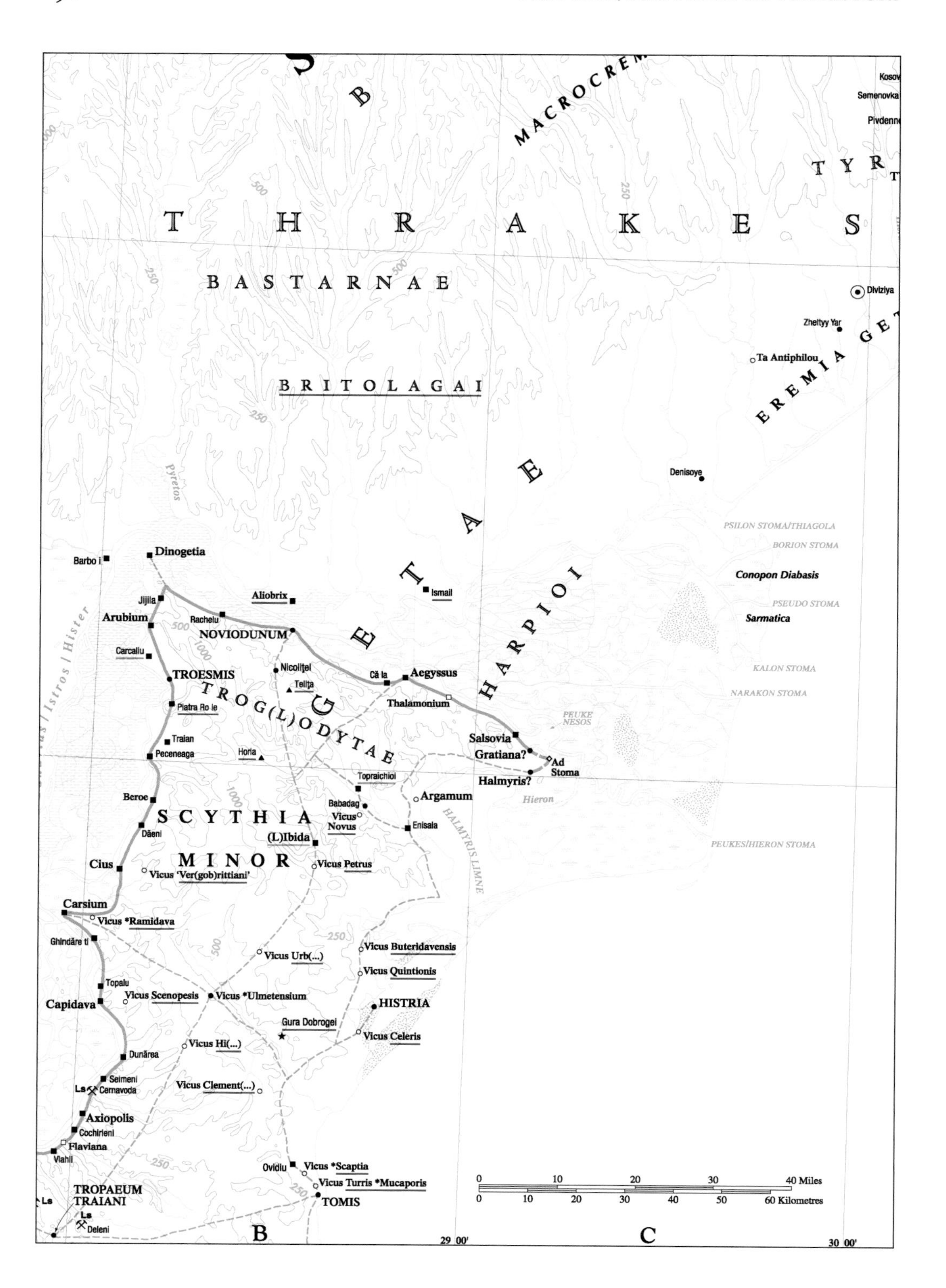

The atlas includes an alphabetical gazetteer of names and a fourteen-hundred-page Map-by-Map Directory on CD–ROM. For each feature mapped in the atlas, the directory provides alternate ancient names and their modern equivalents, periods of occupation, and a bibliography. For each map, the directory also provides a compiler's introduction, pointing out characteristics of the map that call for special explanation. A good example comes from the introduction to Map 90, showing the southwest coast of the Caspian sea, by S. E. Kroll:

> *Research in this region has always been hampered by national boundaries. Most of the area covered by the map belongs to Iran, but the northern parts belong to (former Soviet) Azerbaijan, to Armenia, and to the autonomous but disputed regions of Nakhichevan and Nagorno Karabakh. Apart from reports from travelers, no scientific research was undertaken in this part of Iran before 1945, nor has any comprehensive survey of ancient sites yet been carried out. As a result, many of the sites marked here have been discovered by chance, and not by intensive research. Because of the border situation, almost no topographic research has been possible in Armenia, northern Azerbaijan and the autonomous regions.*[4]

The directory is also available in a separate, two-volume print version.

Designing the maps and directories for the *Barrington Atlas* caused us to think carefully about how to represent historical knowledge in a geographic context. We wanted to organize and communicate detailed information consistently and unambiguously. We also wanted to represent the degree of certainty attached to the information, fine distinctions that are nonetheless critical to historical scholarship for any period because they can affect the nature and reliability of conclusions drawn from subsequent analysis. The symbols, line styles, area patterns, and typographic conventions employed on the maps reflect these concerns, providing information beyond that found in the gazetteer or the directory. For example, each category of site—settlement, monastery, lighthouse, bridge, mine—has its own point symbol. Different kinds of lines distinguish between roads, aqueducts, and canals. Solid lines and symbols indicate certainty about the location of a feature; hollow symbols and dashed lines indicate that the location shown on the map is only approximate. Both the atlas maps and directory employ typographical conventions to communicate varying degrees of certainty about names *(figure 5)*: whether a particular feature known from archaeology should be equated with a particular name known from an ancient source; when a spelling error is suspected in a name provided by an ancient source; or when a name has been reconstructed. Names mentioned in ancient sources that are impossible to map even approximately are listed as unlocated toponyms in the directory. Users who consult the alphabetical gazetteer and the searchable directory on CD–ROM can be confident of finding virtually any geographic name they encounter in an ancient Greek or Roman work. The atlas then affords them the opportunity to assess the confidence with which that place or feature can now be located, and to see it in its setting on the map.

The *Barrington Atlas* is not a GIS, but as GIS technology developed and improved during the 1990s, the Atlas Project used it to produce about 60 percent of the maps. The first maps to be compiled (most of those at the 1:500,000 scale) were produced using the film-based method that had been standard since the 1920s. The cartographers at MapQuest.com used these techniques to prepare transparent film compilation bases, which were sent to the scholars who compiled the maps. They marked changes on transparent Mylar® overlays, which were then reviewed by atlas project editors and then returned to the cartographers for final proofing and production *(figure 6)*.

From 1994, however, maps were prepared using spatial data from the Digital Chart of the World (DCW), which was selected, projected, cropped, and printed on transparent film to prepare the compilation bases. To correct omissions and errors in the data, the cartographers edited it in a GIS using raster scans of the paper charts that were the basis for the DCW.[5] As GIS became more central to the work of the atlas project, the team became more aware of the potential for new work that would use geospatial analysis and other data sources to build upon the compilation and organization efforts associated with the *Barrington Atlas*. Experience with GIS did, in fact, condition some final decisions about the organization of information in the Directories and Gazetteer. It also confirmed the value of the time and effort that had gone into design and systematization of feature types and attribute data.

Although the Atlas Project was commissioned as a finite effort with a tangible conclusion in mind, namely the publication of a printed atlas, those involved came to recognize that publication ought to also represent the beginning of a renewed dialogue

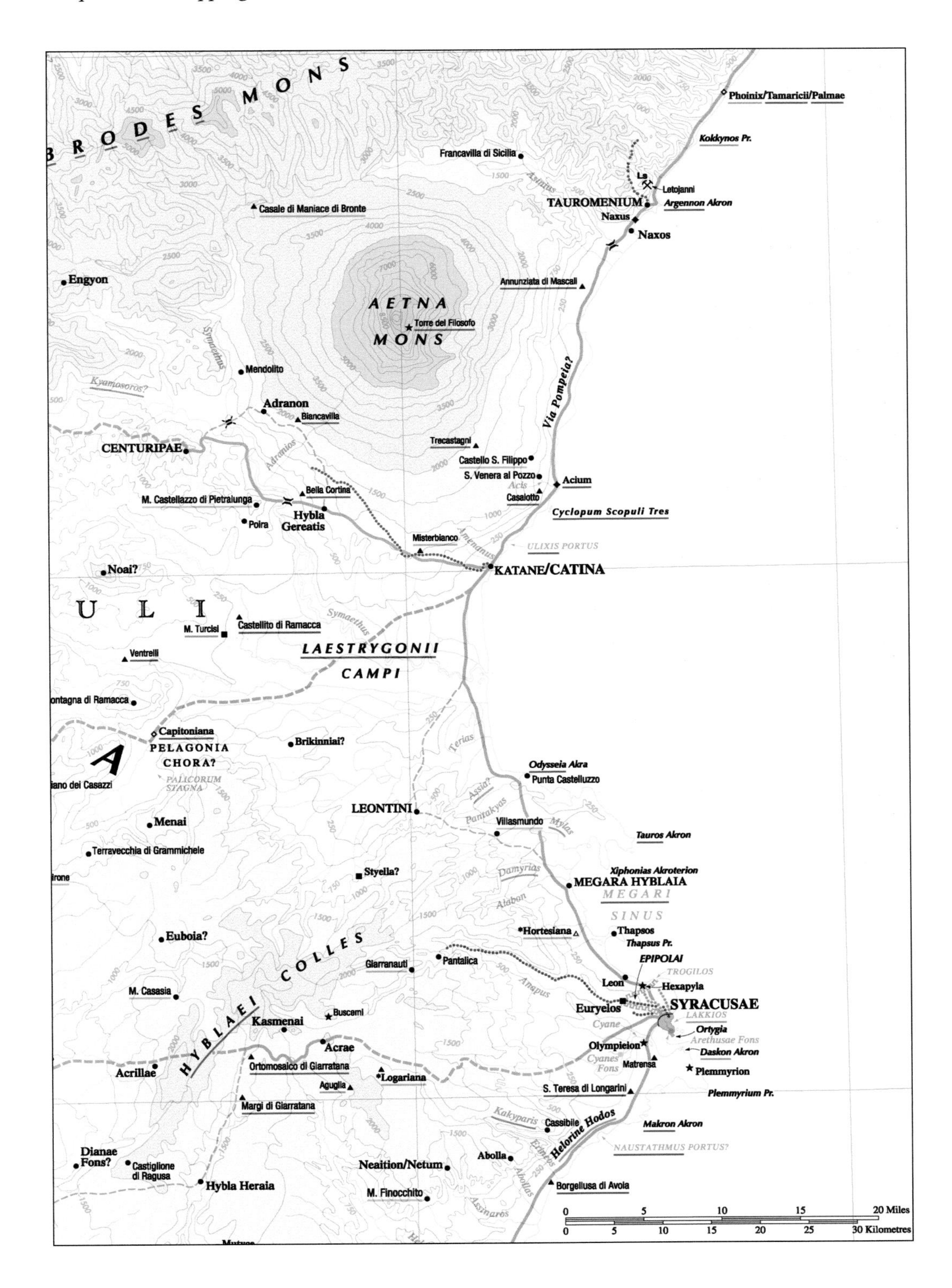

Figure 5. Site certainty on the east coast of Sicily
Sicily played an important role in the economy and politics of Greek and Roman civilization. Even this well-recorded and intensively studied place, however, has archaeological sites that scholars have not yet definitively matched to sites mentioned in ancient sources. There is no question about the location of Mount Etna (ancient Aetna Mons), but question marks indicate the uncertain assignment of names from ancient literature like Styella, Euboia, and Dianae Fons to sites identified by archaeologists. Sites with no known ancient referent are labeled with their modern names, such as "Francavilla di Sicilia" just north of Mount Etna. Red underlining means a site was active only in the early Roman Empire.

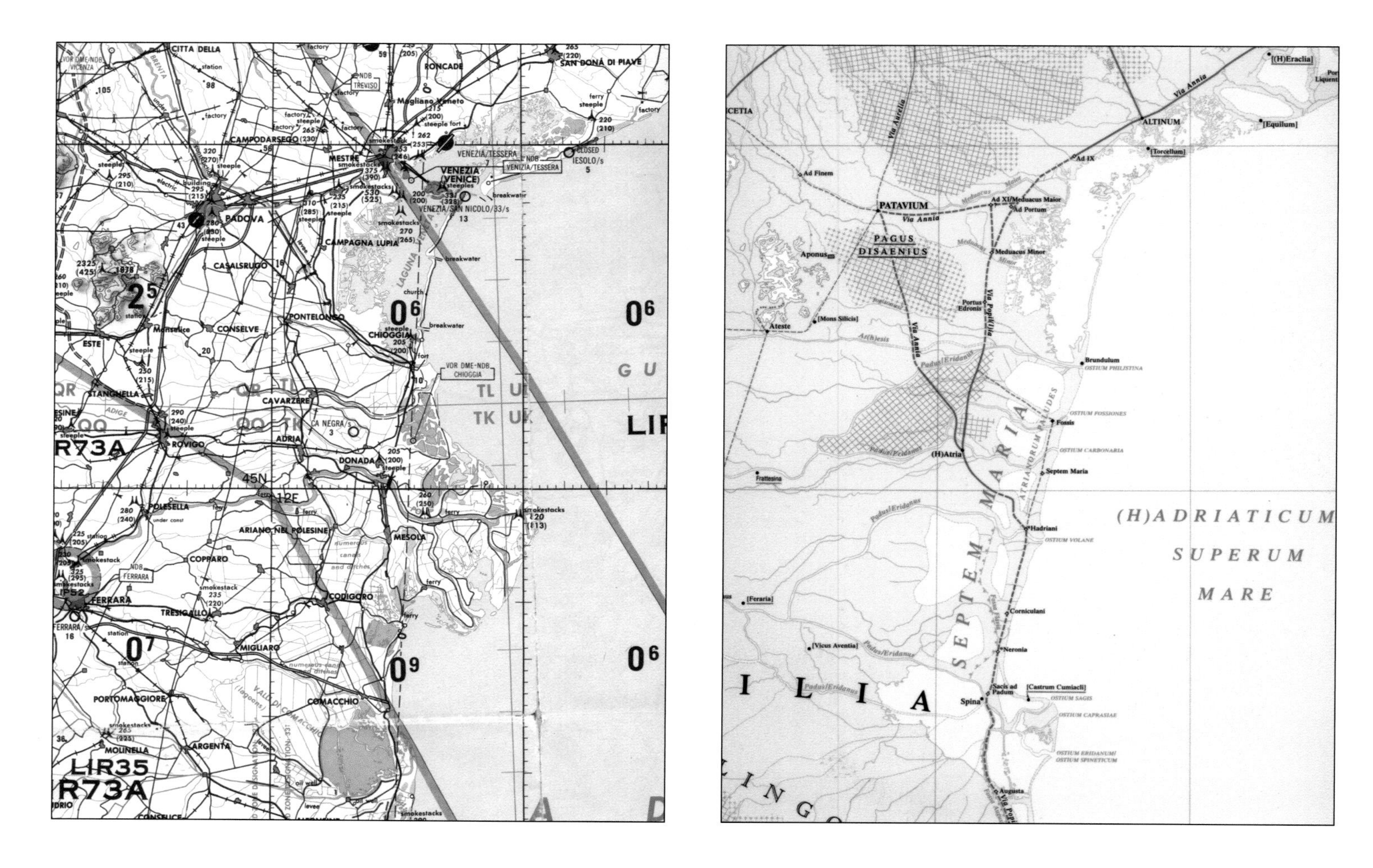

Figure 6. Returning the Po River delta to its ancient appearance

These two maps show the same area of the Po River delta on the northeast coast of Italy. To the left is an excerpt from a Defense Mapping Agency (now the National Imagery and Mapping Agency) tactical pilotage chart, from which cartographers extracted modern coastline, river, and elevation data. Siltation and floods have altered this coastline and coastal settlement patterns since antiquity. The map on the right shows the region as it was configured between 1000 b.c. and a.d. 650, according to archaeological findings and geologic modeling. Dotted lines show aqueducts, solid and dashed lines mark roads. Grid patterns represent centuriation, a distinctively Roman division of land into a regular grid. Branches of the “Padus/Eridanus” (Po River) flow around an area of centuriation just before they enter the “(H)Adriaticum/Superum Mare” (Adriatic Sea).

between historians, classicists, and geographers. This view crystallized as the project neared its end. The preface to the atlas announced the establishment at the University of North Carolina, Chapel Hill, of a permanent Ancient World Mapping Center to promote cartography and geographic information science as essential disciplines within the field of ancient studies. Chief among the goals set for the Center were the creation of a digital atlas together with a georeferenced database of ancient geographical information. It is possible, therefore, to view the completed atlas as the first step in the creation of a historical GIS of the Greek and Roman world *(figure 7).*

The Ancient World Mapping Center is heir to the collected data of the Atlas Project, which its staff has already begun to convert and reformat for incorporation into GIS systems. All of the data contained in the Map-by-Map Directory now resides in a database on a high-performance computing platform. This database captures all the nuances of the original directory, but disentangles their logical, semantic structure from the typographic conventions required by the printed volume. For example, where the directory adds a question mark after a name to indicate uncertainty about whether it should be assigned to that particular place, the database stores this information in two separate fields, one for the name itself, and another to record degree of certainty. Similarly, where the directory uses roman type for ancient names and italic type for modern names, the database contains one field for names and a second to indicate whether the name is ancient or modern.

Organizing the attribute data into discrete categories that can be related without conceptual overlap makes it possible for GIS users to exploit information about each place to answer their own questions. For example, one could plot all sites in a region whose ancient name is certain and contains a linguistic element indicating “safety” or “protection.” From there, one could determine which types of sites (forts perhaps) are statistically most likely to include that root in their name. Further analysis might show that both forts and settlements in particular types of landscape settings are likely to have such a name, but only during a particular period.

The Center is also assessing how a GIS can incorporate the rest of the spatial information and attribute data contained in the maps, including the extensively reconstructed shorelines, drainage networks, and terrain elevations. Initial tests have shown that it takes about twenty hours to

Figure 7. The center of empire This map of ancient Rome (Roma) and its immediate environs is one of three at the scale of 1:150,000. Using a larger scale enables us to show the density of urban and suburban settlement, major and minor roads, and the complex network of aqueducts. Dashed lines indicate roads whose paths can be determined only approximately. Among the other sites of historical interest are the tufa and travertine quarries, marked by a crossed pickax and sledgehammer and the letters "Tu" and "Tr," respectively. Tufa was a sturdy, fire-resistant building stone; travertine was a decorative siding material that was later replaced by marble from elsewhere in the Roman world. Gray fill (e.g., at Bovillae and Villa Domitiani in the lower right corner of the map) indicates the extent of built-up urban areas in major settlements.

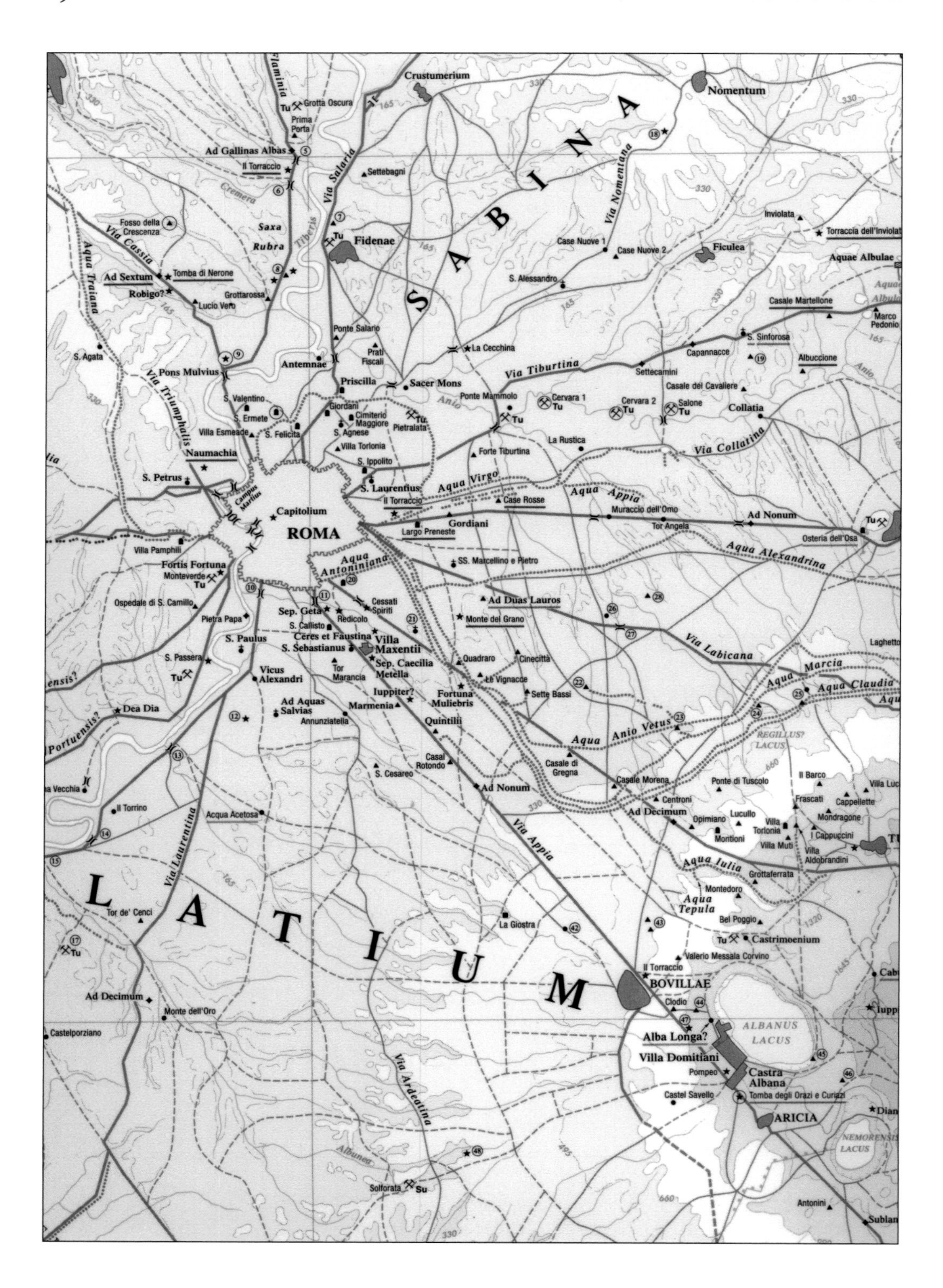

carry out the format translation and registration of points and linework on a typical *Barrington Atlas* map that is already in digital form, plus the time for checking and verification. We are still trying to estimate the more tedious tasks of reassociating label text with registered objects and extracting the attribute data associated with type and symbol styles. Registration and vectorization of the conventionally produced maps is expected to be the most time-consuming task of all.

As Anne Knowles points out in her introduction to this volume, successful historical GIS projects require a heavy investment in data collection and classification combined with tools and techniques for spatial analysis, modeling, and data aggregation. To serve the spatial data and analysis needs of the ancient history and classics communities, we are developing a public, collaborative workplace for the study of ancient geography that will capitalize upon the organized assemblage of data now presented in paper form through the *Barrington Atlas.* This GIS will provide more than just the collaborative visualization and analysis tools currently available in other collaborative GIS environments; it will also facilitate collaborative management, update, and creation of new spatial data by enabling any interested individual to propose changes and additions to the core *Barrington Atlas* data. These public proposals will be vetted through a process of online peer review, in a manner similar to that now employed by the Suda Online (SOL) project.[6] Where SOL facilitates the public translation, editing, and review of entries in an important Byzantine encyclopedia, our GIS (dubbed Pleiades after the daughters of Atlas) will permit students, researchers, and private enthusiasts from a variety of disciplines to suggest changes, additions, and deletions to any of the data, ranging from a single attribute value for a single site to spatial coordinates for a large group of objects. These proposals will be open to critique and improvement by other Pleiades users until they are finally accepted or rejected by the Pleiades editorial board.

Once Pleiades is online, a student reading Demosthenes in the Perseus Digital Library[7] will find instant guidance on the questions of "which Apollonia and why?" *(figure 8).* This guidance will arrive on demand because the geographic visualization tools already built into Perseus and other systems will combine information from digitized historical commentaries with the continuously updated authoritative geographic data Pleiades will provide.

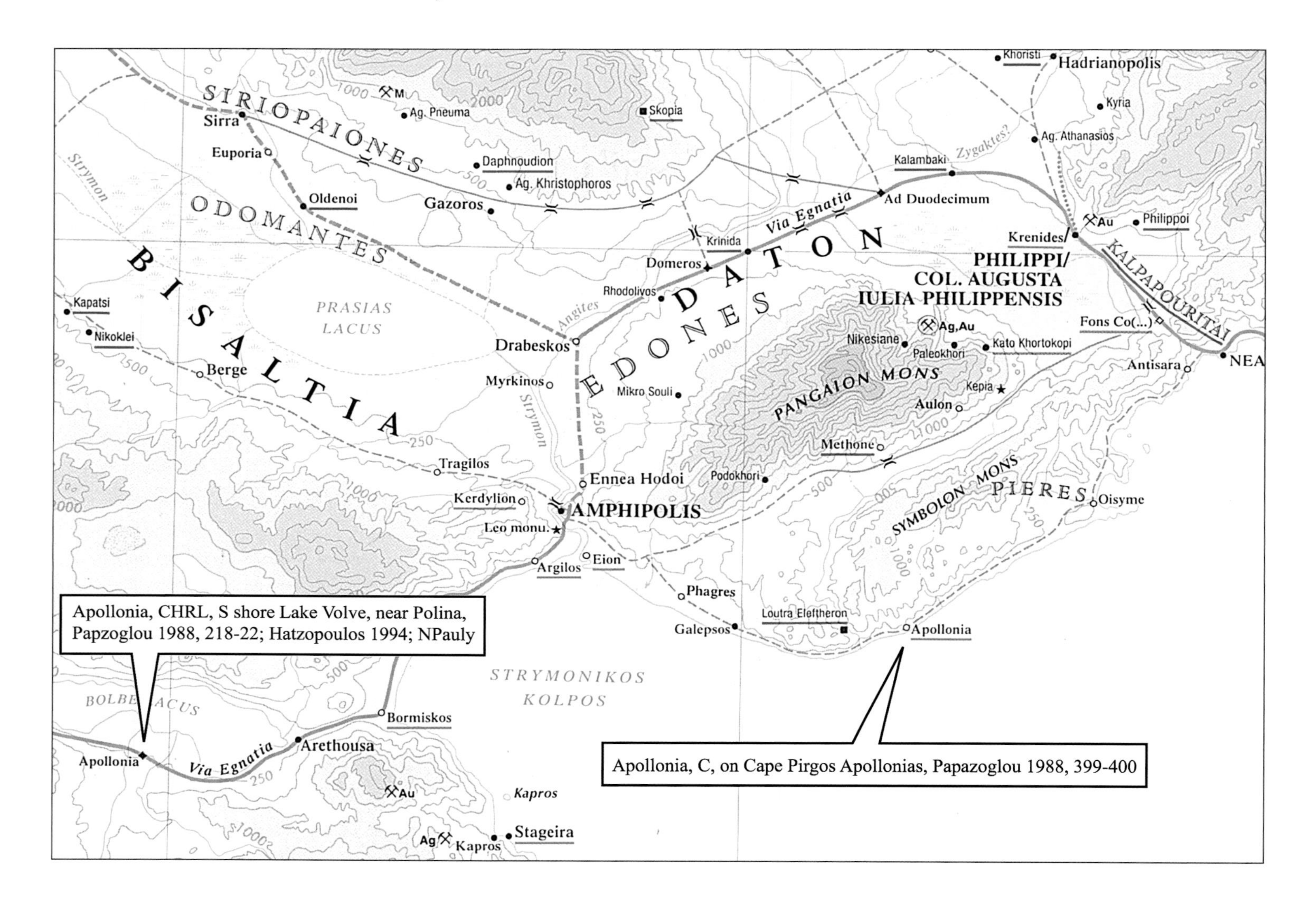

Figure 8. Which Apollonia?

Pleiades will make available all the information contained in the Barrington Atlas and its Map-by-Map Directory. The map detail shown here includes two of the places called Apollonia in ancient Greece. Both were on the coast road (Via Egnatia), one to the west and one to the east of Amphipolis. The diamond symbol for the western Apollonia indicates that it was a road station. The eastern Apollonia, however, was a full-fledged settlement, as indicated by the circular symbol. Its symbol is hollow because we cannot ascertain its precise location. The additional information provided by the directory, contained in callouts in the figure, shows that the western Apollonia was active in the Classical, Hellenistic, Roman Imperial, and Late periods (C, H, R, L following the name). The eastern Apollonia was occupied only in the Classical period. Because Demosthenes was writing at the time of transition between the Classical and Hellenistic periods, we cannot ascertain from the period information which Apollonia he was discussing. We will have to turn to the bibliography cited in the directory entry to learn that he was speaking of the city of Apollonia, not the road station. Pleiades will incorporate all this information, allowing users to quickly produce maps or select data that depicts only a particular period or a particular type of place. It will also permit users of the system to add bibliographic information as new work is published.

Our vision also extends to the cultural historian who will use the Pleiades system to find and analyze the data needed to reassess Roman occupation of the Po valley. Pleiades will give this person access to a digital elevation model derived from Shuttle RADAR topography data and digitized World War II aerial photographs from the ADEPT collection.[8] A link to the Electronic Cultural Atlas Initiative (see chapter 12) would bring up a GIS data set assembled by a European research team, showing patterns of landholding in the Po valley during the Middle Ages. Pleiades will also provide, from its own database, a GIS data set locating known Roman-era settlements in the valley, together with names and an up-to-date bibliography for each.

Thanks to the happy convergence at this moment of mature GIS technologies, a high-quality data set already moving to digital format, and an increasingly interconnected world, there is at last the opportunity to redress the lack of a strong spatial component in the study of ancient history. The other contributions to this volume make it abundantly clear that the Ancient World Mapping Center is neither alone in sensing such opportunities, nor without seasoned guides and companions for the journey.

Map credits

Figure 1
Detail from Map 26: Peloponnesus, drawn by C. Müller and engraved by S. Jacobs and I. Dalmont, in W. Smith and G. Grove, eds., *An Atlas of Ancient Geography, Biblical and Classical* (Boston: Little, Brown and Co., 1874).

Figure 2
Barrington Atlas Locator Outline Map. Copyright Princeton University Press 2000.

Figure 3
Detail from Barrington Atlas Map 3: Asia Occidentalis. Compiled by M. Roaf and the Project Office, 1998. Copyright Princeton University Press 2000.

Figure 4
Detail from Barrington Atlas Map 23: Tomis-Olbia-Chersonesos. Compiled by D. Braund, 1995. Copyright Princeton University Press 2000.

Figure 5
Detail from Barrington Atlas Map 47: Sicilia. Compiled by R. J. A. Wilson, 1997. Copyright Princeton University Press 2000.

Figure 6
6a. Detail from TPC F-2B, Defense Mapping Agency, 1968.

6b. Detail from Barrington Atlas Map 40: Patavium. Compiled by M. Pearce, R. Peretto, and P. Tozzi, 1994. Copyright Princeton University Press 2000.

Figure 7
Detail from Barrington Atlas Map 43: Latium Vetus. Compiled by L. Quilici and S. Quilici Gigli, 1995. Copyright Princeton University Press 2000.

Figure 8
Detail from Barrington Atlas Map 51: Thracia and its Map-By-Map Directory. Compiled by E. N. Borza, 1994. Copyright Princeton University Press 2000.

Further reading

Boettcher, R. "Collaborative GIS in a Distributed Work Environment." M. Eng. thesis, University of New Brunswick, 1999.

Chavez, R. "Using GIS in an Integrated Digital Library." *Proceedings of the 5th Annual ACM Digital Library Conference,* 2000: 250–51.

MacEachren, A. "Cartography and GIS: Facilitating Collaboration." *Progress in Human Geography* 24, no. 4 (September 2000): 445–56.

———. "Cartography and GIS: Extending Collaborative Tools to Support Virtual Teams." *Progress in Human Geography* 25, no. 3 (September 2001): 431–44.

Rosenstock, B., and M. Gertz. "Web-based Scholarship: Annotating the Digital Library." *Proceedings of First ACM/IEEE-CS Joint Conference on Digital Libraries,* June 24–28, 2001: 104–5.

Smith, D., and G. Crane. "Disambiguating Geographic Names in a Historical Digital Library." *5th European Conference on Digital Libraries,* September 2001: 127–36.

Spasser, M. *Computational Workspace Coordination: Design-in-use of Collaborative Publishing Services for Computer-mediated Collaborative Publishing.* Ann Arbor: University of Michigan Press, 1999.

Tooby, "Re-inventing Scholarly Communication," *EnVision* 17, no. 1 (January–March 2001). Available online at www.npaci.edu/enVision/v17.1/re_inventing.html.

Witten, I., D. Bainbridge, and S. Boddie. "Power to the People: End-User Building of Digital Library Collections." *Proceedings of First ACM/IEEE-CS Joint Conference on Digital Libraries,* June 24–28, 2001: 94–103.

Online resource
Hill, L., et al., *Alexandria Digital Library Gazetteer Development Information,* Alexandria Digital Library Project: alexandria.sdc.ucsb.edu/~lhill/adlgaz

Notes

1. *On Halonnesus,* 7.28; Loeb translation by J. H. Vince.
2. The development of the division between classics and geography/cartography, and the various attempts to redress it, are traced in Richard J. A. Talbert, "Mapping the Classical World: Major Atlases and Map Series 1872–1990," *Journal of Roman Archaeology* 5 (1992): 5–38.
3. R. S. Bagnall, ed., *Research Tools for the Classics: The Report of the American Philological Association's Ad Hoc Committee on Basic Research Tools* (Chico, Calif., 1980), 27.
4. Richard J. A. Talbert, ed., *Map-by-Map Directory to Accompany the Barrington Atlas of the Greek and Roman World* (Princeton, N.J.: Princeton University Press, 2000), 1,292.

Continued

Notes (continued)

5. Compilers continued to prepare Mylar overlays from the printed film. When the compilation overlays arrived at MapQuest.com, the cartographers scanned the Mylar and film compilations, reregistered them with the basemap materials, and moved all the layers to a graphic editing program, where the compilers' additions were vectorized, deletions and changes made to the base layers, and styles and typography applied. GIS processing was also used to produce the tinted, shaded relief of the six introductory maps.
6. Suda Online: www.stoa.org/sol
7. Perseus Digital Library: www.perseus.tufts.edu
8. Alexandria Digital Earth Prototype: www.alexandria.ucsb.edu

CHAPTER TWELVE

THE ELECTRONIC CULTURAL ATLAS INITIATIVE AND THE NORTH AMERICAN RELIGION ATLAS

Lewis R. Lancaster and David J. Bodenhamer

In 1997, sixteen scholars met at the University of California, Berkeley, to discuss the problems of using digital technology in cultural studies. They identified a number of issues, including the difficulties of comparing and combining data sets that use different standards and operating systems; scholars' problems in acquiring necessary technical expertise; the ephemeral nature of online data; institutions' tendency to undervalue electronic publications; and the need for a common system for organizing and searching digital resources. The group's response to those issues was to form the Electronic Cultural Atlas Initiative (ECAI)[1], an association whose participants now include more than six hundred scholars, librarians, and information technology experts from around the world.

From ECAI's inception, its members recognized the importance of both the temporal and spatial dimensions of cultural studies. Participants in the first meeting were drawn together by their interest in studying events that spanned centuries and continents, such as the gradual spread of Buddhism from India throughout Asia. Despite their geographical interests, however, few of ECAI's early members had heard of GIS or considered location to be a key for organizing and linking complex historical data sets. At the second meeting later in 1997, Professor Larry Crissman of Griffith University in Australia persuaded the group that GIS should become the standard format for spatial data in ECAI-supported projects.

Participants also agreed that all data included in the ECAI online "atlas" (a searchable collection of historical data sets) should be referenced geographically so that it could be used with GIS. At the very least, this means that each project's geographical study area is identified. In some cases, each image in the data set has a separate Clearinghouse entry to assist searching. ECAI has begun to develop tools that will help participants automatically mark the geography of their material using online *gazetteers.* One such project is being done in coordination with Academia Sinica in Taiwan, where the forty million characters of the Dynastic History of China project can be searched using an online gazetteer of place-names from the historical atlases. Since the place-names have a latitude and longitude field in the gazetteer, this tool will allow us to map all sites mentioned in the Dynastic Histories. It will in turn allow scholars to move from the text to a map that provides linkage to any data registered for the particular spot.

Thus GIS has come to play a central role in the overarching goal of ECAI—namely, to facilitate digital research by collecting, standardizing, and cataloging online databases and projects. The close relationship between those three activities is evident in the ECAI Metadata Clearinghouse System. Like a library, the Clearinghouse catalogs resources and applies standards for their identification and retrieval, including geographical and historical information. Unlike traditional libraries, however, the Clearinghouse server and browser distribute information about digital resources to users wherever they may be. Users can also use the Clearinghouse metadata editor to conform their own work to ECAI standards.

The aim of facilitating use of digital technologies has spawned many other activities within ECAI's global network. The ECAI Technical Group, based at Berkeley, develops database architecture and identifies emerging needs for infrastructure and software in the academic community. International meetings, convened in a different location every six months, bring together growing numbers of affiliates to present their research, discuss their uses of digital technology, and learn new methods from ECAI trainers. Week-long ECAI Institutes, cosponsored by The Polis Center at Indiana University–Purdue University, Indianapolis, provide intensive training for small groups of scholars, librarians, and administrators who need to get up to speed quickly in information technologies relevant to scholarship in the humanities and social sciences.

To foster interdisciplinary scholarship, ECAI provides support to teams of scholars who are collaborating on digital projects. Collaboration is new for most ECAI affiliates, who come predominantly from the humanities. Their customary modes of solitary scholarship are often rendered impossible, or at least impractical, by the demands of digital technology. ECAI has also formed partnerships with other academic institutions that share its aims of developing standards for digital data, information technology architecture, and Internet publication. The original vision of enabling scholars to analyze change over long periods of time led to ECAI's ongoing collaboration with Professor Roland Fletcher and Dr. Ian Johnson at the Archaeological Computing Laboratory of the University of Sydney. They are developing a GIS program called TimeMap, which is now the ECAI standard for temporal GIS.

The TimeMap software allows users to open the browser and choose data that is registered in the ECAI Metadata Clearinghouse. The Clearinghouse indexes data sets accessible via the Internet and provides two sorts of metadata: descriptive categories and connection metadata about the type of server, naming of tables and fields, passwords, georeferencing, and other information users might need to use a data set. In response to user queries, the Clearinghouse delivers data in displays like that shown in figure 1. In

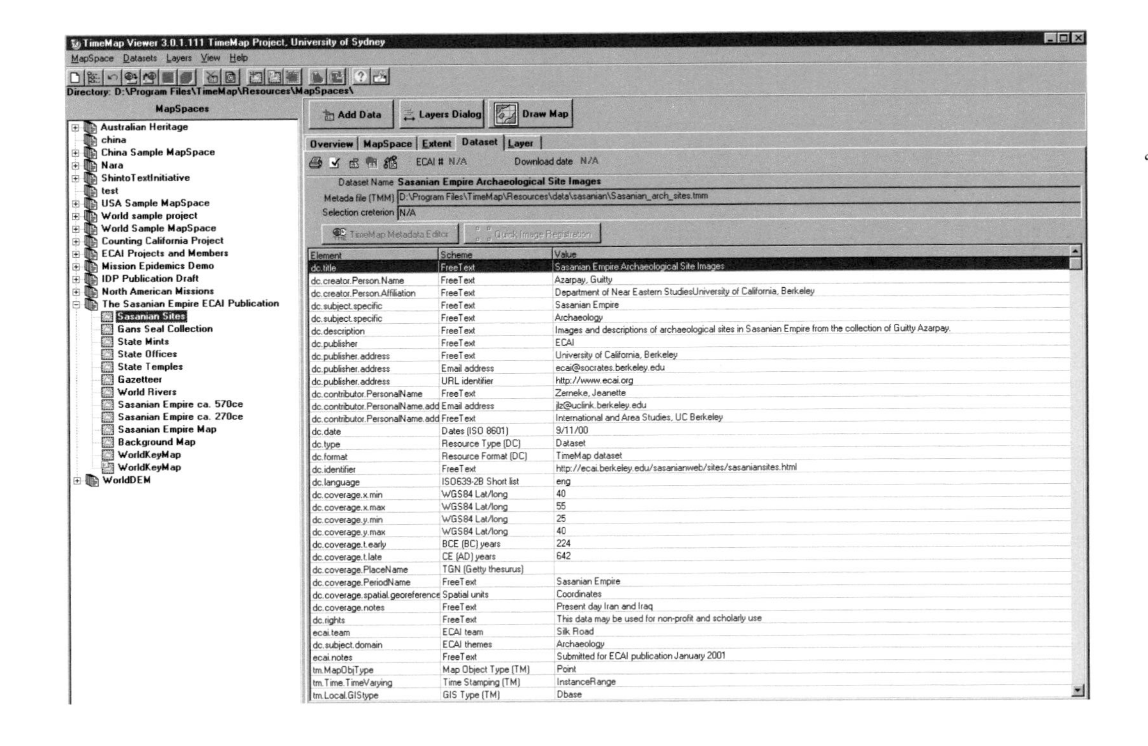

Element	Scheme	Value
dc.title	FreeText	Sasanian Empire Archaeological Site Images
dc.creator.Person.Name	FreeText	Azarpay, Guitty
dc.creator.Person.Affiliation	FreeText	Department of Near Eastern StudiesUniversity of California, Berkeley
dc.subject.specific	FreeText	Sasanian Empire
dc.subject.specific	FreeText	Archaeology
dc.description	FreeText	Images and descriptions of archaeological sites in Sasanian Empire from the collection of Guitty Azarpay.
dc.publisher	FreeText	ECAI
dc.publisher.address	FreeText	University of California, Berkeley
dc.publisher.address	Email address	ecai@socrates.berkeley.edu
dc.publisher.address	URL identifier	http://www.ecai.org
dc.contributor.PersonalName	FreeText	Zerneke, Jeanette
dc.contributor.PersonalName.add	Email address	jlz@uclink.berkeley.edu
dc.contributor.PersonalName.add	FreeText	International and Area Studies, UC Berkeley
dc.date	Dates (ISO 8601)	9/11/00
dc.type	Resource Type (DC)	Dataset
dc.format	Resource Format (DC)	TimeMap dataset
dc.identifier	FreeText	http://ecai.berkeley.edu/sasanianweb/sites/sasaniansites.html
dc.language	ISO639-2B Short list	eng
dc.coverage.x.min	WGS84 Lat/long	40
dc.coverage.x.max	WGS84 Lat/long	55
dc.coverage.y.min	WGS84 Lat/long	25
dc.coverage.y.max	WGS84 Lat/long	40
dc.coverage.t.early	BCE (BC) years	224
dc.coverage.t.late	CE (AD) years	642
dc.coverage.PlaceName	TGN (Getty thesurus)	
dc.coverage.PeriodName	FreeText	Sasanian Empire
dc.coverage.spatial.georeference	Spatial units	Coordinates
dc.coverage.notes	FreeText	Present day Iran and Iraq
dc.rights	FreeText	This data may be used for non-profit and scholarly use
ecai.team	ECAI team	Silk Road
dc.subject.domain	ECAI themes	Archaeology
ecai.notes	FreeText	Submitted for ECAI publication January 2001
tm.MapObjType	Map Object Type (TM)	Point
tm.Time.TimeVarying	Time Stamping (TM)	InstanceRange
tm.Local.GIStype	GIS Type (TM)	Dbase

FIGURE 1. ECAI WEB PUBLICATIONS
THE OPENING SCREEN OF THE TIMEMAP VIEWER DISPLAYS THE LIST OF ECAI WEB PUBLICATIONS (UNDER "MAPSPACES") AVAILABLE THROUGH THE ECAI CLEARINGHOUSE. WHEN THE USER SELECTS A PUBLICATION—HERE, THE SASANIAN EMPIRE—TIMEMAP BRINGS UP THE METADATA AS IT HAS BEEN CATALOGED ACCORDING TO THE DUBLIN CORE STANDARD. THE DUBLIN CORE IS BECOMING THE INDEXING STANDARD FOR LIBRARIES IN NORTH AMERICA AND AROUND THE WORLD. ECAI'S UNIQUE ADDITION TO THE STANDARD IS REQUIRING EACH OF ITS WEB PUBLICATIONS TO INCLUDE FULL METADATA ON GEOGRAPHICAL LOCATION AND TIME.

addition to the usual functions of a GIS, TimeMap's time scroll bar enables users to set the period of time they want displayed. Through links to the ECAI Clearinghouse, users can also search for related data sets.

One of ECAI's first TimeMap projects was Guitty Azarpay's Web publication on the Sasanian seal collection of Edward Gans, which is housed at the University of California, Berkeley *(figures 2–5)*. The Sasanian Empire (224–642 C.E.) was founded by the last great Persian monarchy before the Arab conquest of Western Asia in the seventh century. Third-century Sasanian kings gained control over territory that stretched nearly two thousand miles, from the Persian Gulf to the Black Sea and from the Caucasus to modern Afghanistan. One of the few kinds of artifacts to survive from the Sasanian Empire are the engraved gems and stones whose imprint on clay sealed legal and commercial documents. Though most of the documents long since turned to dust, the seals' intricate carvings of royal insignia and religious symbols provide important evidence of Sasanian culture.

The Sasanian Seal Project exemplifies ECAI's work to promote Internet publication of data resources. In partnership with the California Digital Library and other libraries around the world, ECAI is developing standards for publishing and maintaining data and scholarly work on the Internet. By providing a permanent base for such publications, as well as guidelines for how to evaluate them, we hope that scholars' electronic publications will receive the recognition they deserve in hiring, tenure, and promotion decisions. GIS poses special problems for Internet publication because spatial databases are

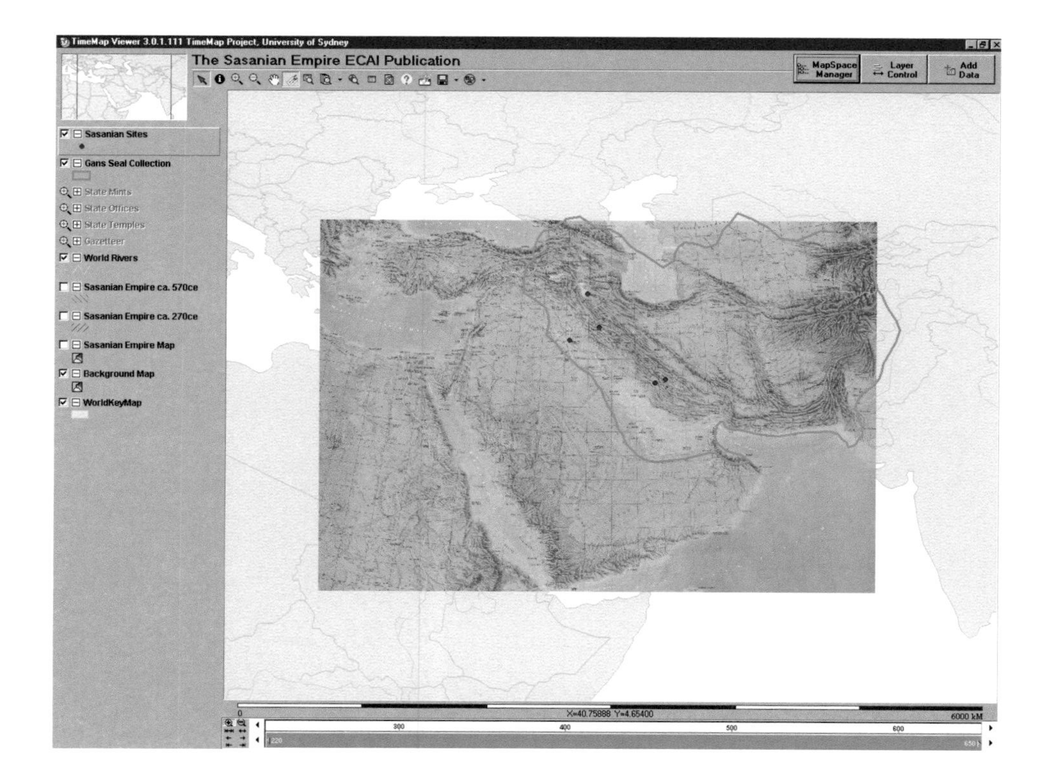

Figure 2. The Sasanian Empire
An overview map of the Sasanian Empire ECAI publication. It shows the topography of the region and the approximate geographical extent of the empire, outlined in blue. Below the map, a time bar (in turquoise) shows the time period of the empire, 220–650 C.E.

more difficult and costly to migrate than textual or numerical databases. GIS can only become a regular part of Internet publication if spatial data is standardized for migration over the long term.

Location is also far from a simple matter when it becomes part of the structure of information technology. Like the scholars involved with the Ancient World Mapping Center (see chapter 11), ECAI collaborative teams have realized the importance of having electronic gazetteers that can locate places according to any of their various historic and linguistic identities. For GIS to become central to historical research, scholars must be able to search digital libraries on place-names that are correctly linked to relevant resources. Another challenge for the use of GIS in the humanities is the imprecision of many locations. Unlike many GIS users, scholars in the humanities must struggle with the fact that empires of the past had, at best, fuzzy boundaries. Even the common terminology of GIS breaks down in the context of many humanistic studies. Drawing the "catchment area" around a religious shrine, for example, requires lines defining a polygon, when both the term and the shape imply far more certainty than the vaguely known but significant regional pattern that the scholar wishes to convey. Mapping the sacred space around a holy site is an even more elusive exercise.

Despite these difficulties, some of the most ambitious GIS-based projects housed within ECAI examine the history of religion. The largest such study to use biographical data is the Hartwell Chinese Officials Project, located at Harvard University, under Professor Peter Bol. The database has more than twenty-five thousand

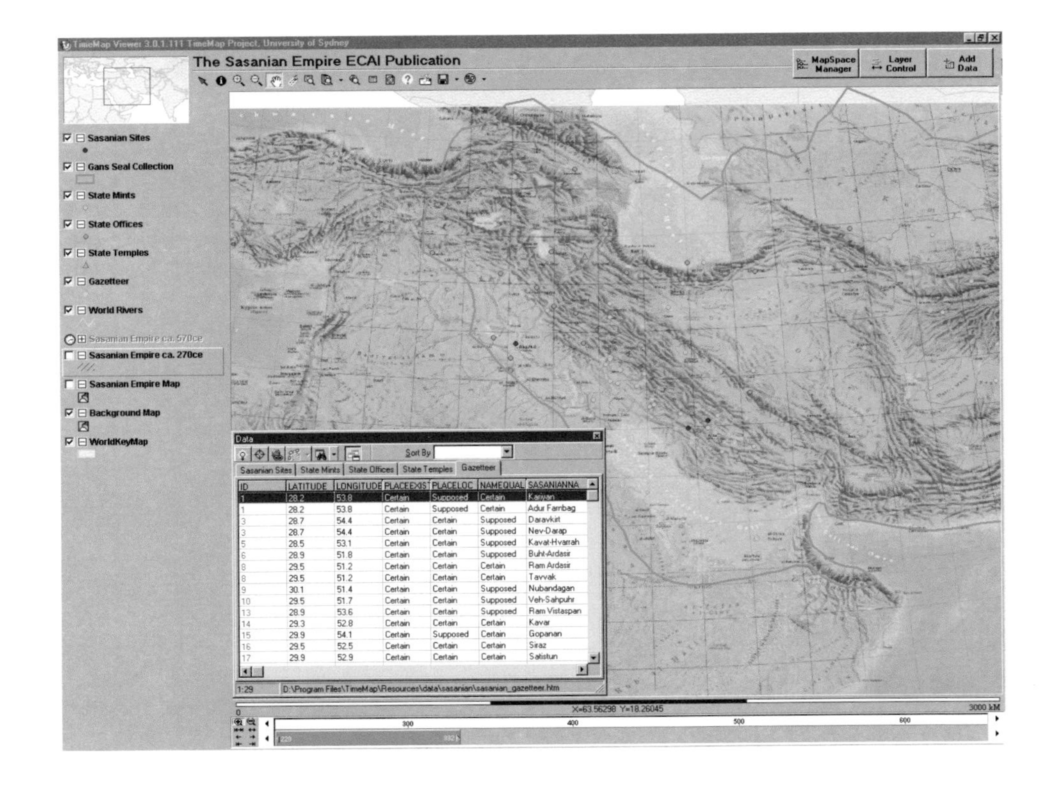

Figure 3. Functions of TimeMap
The legend to the left of the map shows the time and space layers included in the publication. The small inset table below the map shows attributes of locations in the Sasanian Empire whose geographical locations are known, represented by gold dots on the map. The time bar has been set at 220–339 C.E. TimeMap will not display layers whose data falls outside the specified time range, as indicated by the red clock symbol and grayed lettering for the layer "Sasanian Empire ca. 570ce."

names of Chinese officials, covering seven centuries. Professor Robert Hartwell, who collected the original data two decades ago, referenced all of his biographical entries with latitude and longitude: where each official was born, taught, served, died, and was buried. When this material is put into GIS, we can ask such questions as: "Where were court officials in 1445 born?" "Where did they go to school?" "Where did they serve?" Once the software has drawn maps in answer to these queries, we can ask the same questions of other dates and begin to see how power shifted between provinces over the centuries.

In another project, Lionel Rothkrug identified the locations of more than one thousand German pilgrimage shrines from before the Reformation. By combining the GIS map of shrines with layers showing the regional concentration of Catholics and Protestants in later centuries, Rothkrug found that places with pre-Reformation shrines remained Catholic. But a surprise was in store for the researchers. At the GIS Center on the Berkeley campus, John Radke and Rain Simar were viewing the map of shrines when a scholar from Germany walked into the room. Asked to guess what the points meant, he identified

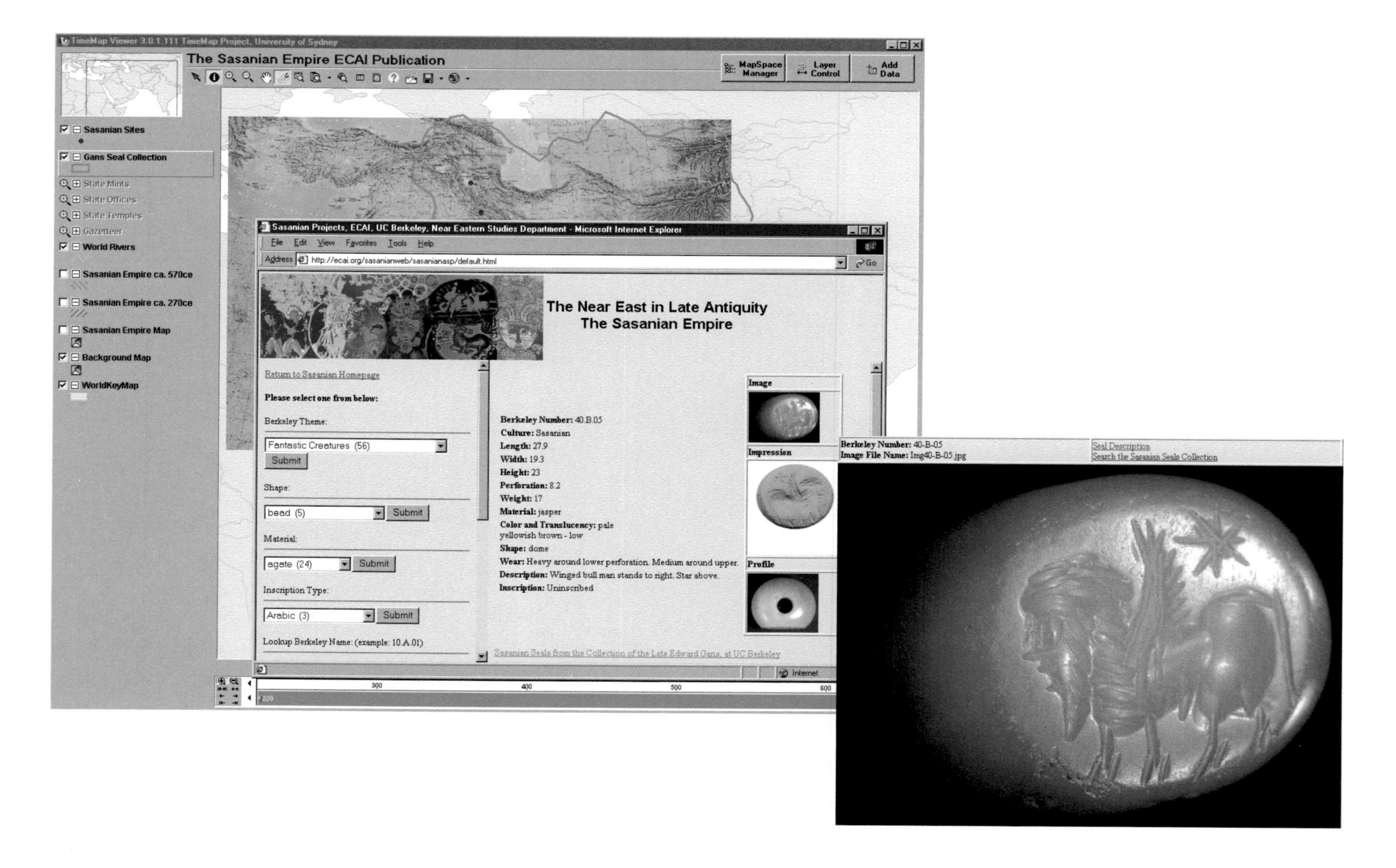

Figure 4. Sasanian seals
Before ECAI published the Gans Collection, few researchers had access to Sasanian seals. Even those who did required magnifying lenses to study the seals' detailed carvings, for the largest seals are less than one inch in diameter. TimeMap can display each seal at many times its original size, alongside information about the seal and a clay impression of the carving, which is often easier to read than the carving itself.

areas with clusters of points as the wealthiest parts of present-day Germany and concluded that the points represented economic data. He was astonished to learn that they were in fact religious sites from many centuries ago. Everyone agreed that seeing unanticipated spatial patterns can shake one's assumptions and prompt intriguing questions for further study.

One of the first ECAI projects to use GIS to disseminate geographically referenced source material is the North American Religion Atlas. The Atlas is being produced at The Polis Center, whose director, David J. Bodenhamer, is group leader of ECAI's North American Team.[2] In its recently completed first phase, the Atlas makes available the demographic data collected by the U.S. Census of Religious Bodies in 1906, 1916, 1926, and 1936 and the decennial censuses conducted since 1952 by the Glenmary Research Center (table 1). Both sources list denominational membership for each county and state of the United States and a host of data on church budgets, facilities, baptisms and confirmations, income, and the like.[3] While they are most complete for mainstream Protestant denominations, the censuses of religion are unparalleled

Figure 5. Architectural context of empire
Only a few architectural ruins mark the site of Ctesiphon, the former capital of the Sasanian Empire, located near present-day Baghdad in Iraq. ECAI publications use the layering and linking functions of GIS to contextualize scholars' research. Here, users can see the physical remains of a powerful, sophisticated society that once controlled much of what is now the Arab world, but from which few artifacts but the seals have survived.

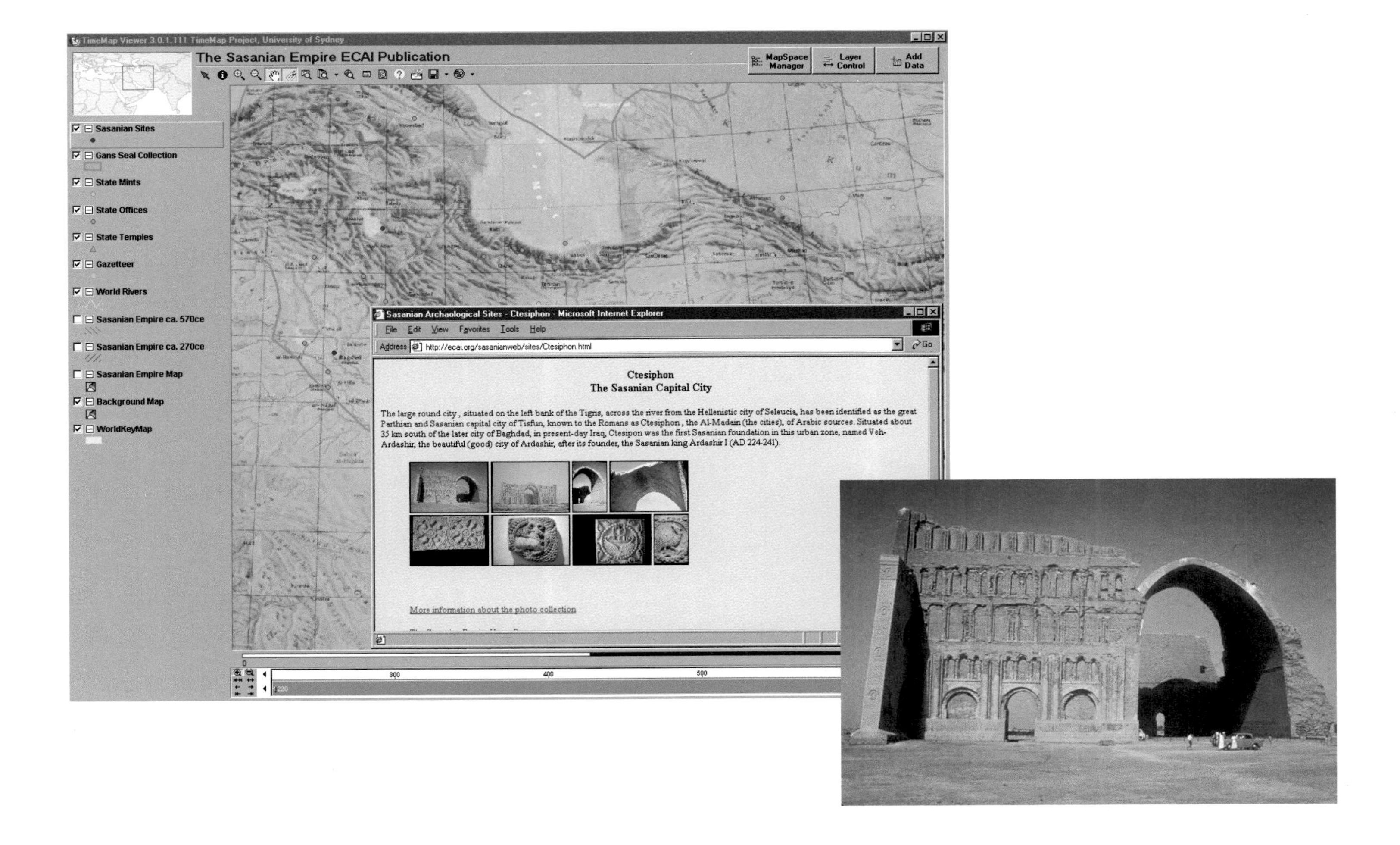

resources for students of all religions in twentieth-century America.

While many scholars have mined these sources, only two works—Edwin Gaustad's *Historical Atlas of American Religion*[4] and William H. Newman and Peter L. Halvorson's *Atlas of American Religion: The Denominational Era, 1776–1990*[5]—extensively map the information they contain. The maps in both works are very useful snapshots of the changing geography of religious denominations and adherence, but they by no means answer all the questions one might ask of the data. The North American Religion Atlas compiles the census data and a wealth of other scholarly resources within a GIS interface that enables researchers to create their own maps—even their own atlases—of America's religious history. The site also houses bibliographic and curricular resources, a glossary of religious terms, and multimedia, including prepared presentations and streamed video and audio *(figure 6).*

The North American Religion Atlas meets ECAI's goal of providing resources for scholars to compare their findings with other studies in the same geographical area by using GIS to locate, layer, integrate, and compare data from various sources.

Table 1. Data layers in the North American Religion Atlas

Layer (1)	Period covered	Scale	Source
U.S. county boundaries	1900–2000 (by decade)	County	USGS
Churches and members	1951	County	Glenmary
Churches, adherents, (2) and members	1971, 1980	County	Glenmary
Denominational membership	1906, 1916,1926, 1936	County	ICPSR
Colonial and Indian missions (economic data, epidemics, names of missionaries)	1500s–1700s	Locality	Data set developed by John Corrigan (3)
Population	1790–2000	County	U.S. Census Bureau

(1) The following variables appear in one or more of the CRB and Glenmary censuses, all available at the county level: adherents by denomination by age (adults and children), gender, race; places of worship and seating capacity; expenditures and debt; Sunday schools and other programs; church property and parsonages; clergy salaries and status (full-time, bivocational); immigrant communications and languages; support of missions and philanthropic institutions; numbers of religiously affiliated hospitals and patients, and schools and pupils.

(2) Adherents is a more expansive category than members and includes persons born into a faith tradition and claimed by it, even though the individuals themselves may not actually be members of an organized congregation.

(3) Corrigan is Edwin Scott Gaustad Professor of Religion and Professor of History, Florida State University.

Through its link to ECAI's Metadata Clearinghouse, scholars will be able to locate and compare data from various studies, using the data resources of the North American Religion Atlas as an analytical framework.

To test the functionality of the Atlas, NARA researchers have developed case studies of religion in Indianapolis and Indiana that draw on the various kinds of data and geographic resources within the GIS. Second Presbyterian Church is the largest and most prestigious congregation in Indianapolis. Its membership historically has boasted many civic leaders, from mayors to industrialists. Plotting the addresses of members across time mirrors both the dynamics of twentieth-century metropolitan growth and the residential migration that led the church to relocate from its historic downtown location to a new home in the suburbs *(figure 7)*. Portraying this movement against the socioeconomic backdrop of the U.S. Census makes it clear that the church continues to represent the city's elite as it has done for generations.

NARA's interactive mapping functions build on the community information system that The Polis Center developed

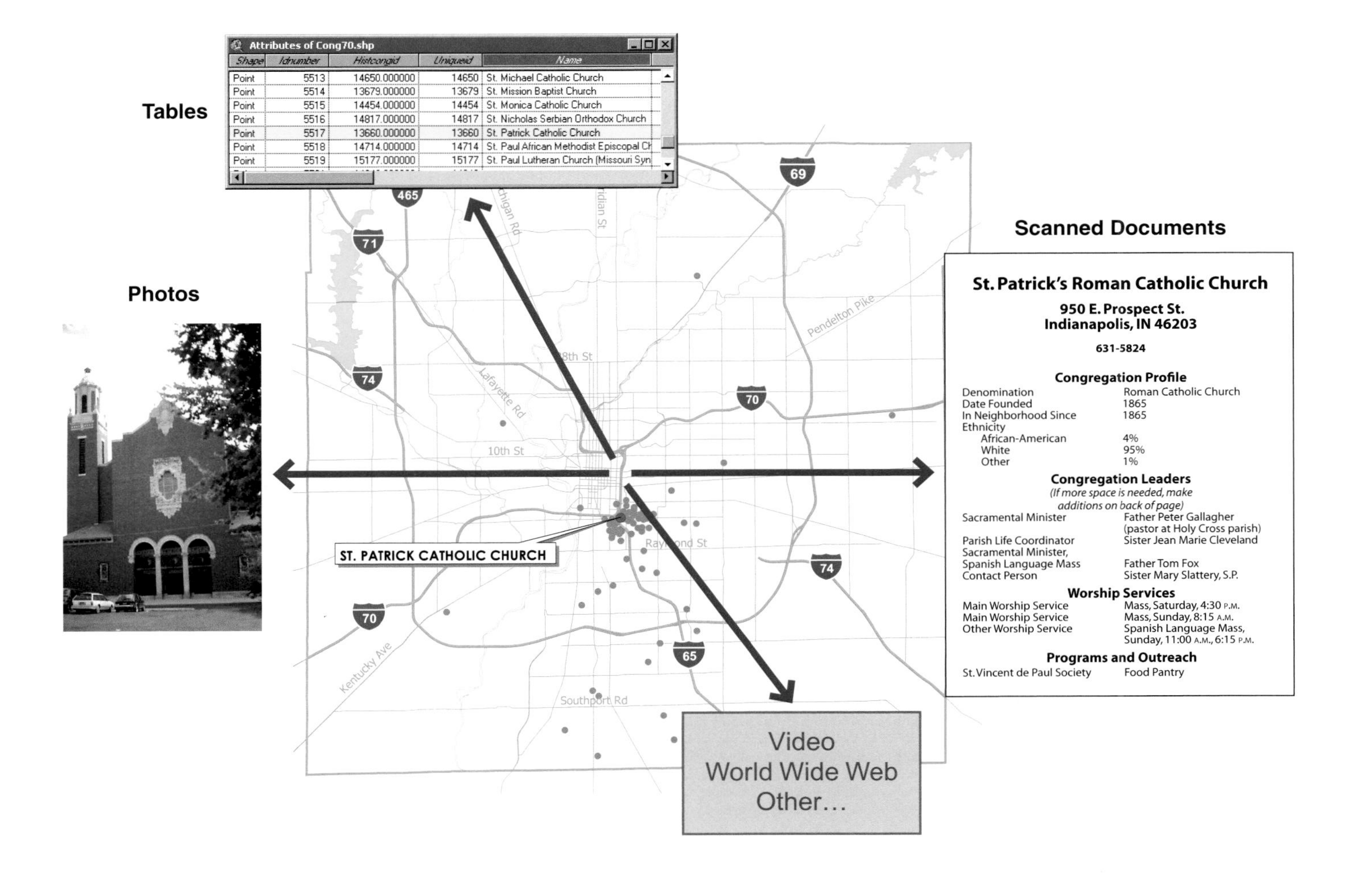

FIGURE 6. GIS IN A MULTIMEDIA ENVIRONMENT
THE NORTH AMERICAN RELIGION ATLAS BRINGS TOGETHER MAPS, PHOTOGRAPHS, TEXT, STATISTICAL DATABASES, AND OTHER RESOURCES ON THE HISTORY OF RELIGION. ITS GIS MAPPING FUNCTIONS ALSO ENABLE USERS TO MAKE THEIR OWN MAPS FROM ATLAS RESOURCES.

Figure 7. Presbyterian Church membership and income in Indianapolis

Between 1950 and 1990, the membership of the Second Presbyterian Church shifted residence from the center of Indianapolis toward the far north side and well into adjacent Hamilton County. Adding census data on median income to the map helps contextualize the migration, the construction of a new church in the suburbs, and the continued affluence of church members.

for the Indianapolis Metropolitan Statistical Area.[6] The Atlas's mapping tool offers a range of options for selecting and displaying data. The query tools allow users to work with data analytically *(figure 8)*. Simple queries locate features by their attributes. Spatial queries are more complex searches, often involving multiple variables. The Atlas also normalizes this data by population and as a percent of total adherents. Mapping variables by the available census years gives the analyst a view of change over time. In future versions, viewers will have the ability to display multiple maps simultaneously and perhaps to animate the data as well.

The history of religion can benefit in various ways from geographical analysis. Mapping the travels of religious leaders and the locations where their adherents became concentrated, for instance, can help explain the diffusion of religious ideas. One can often better understand the symbolic significance of religious architecture by examining it within the context of its physical setting, the geographical distribution of church members, their ethnicity and race, and the proximity and relative size of other religious groups in the area. Perhaps the

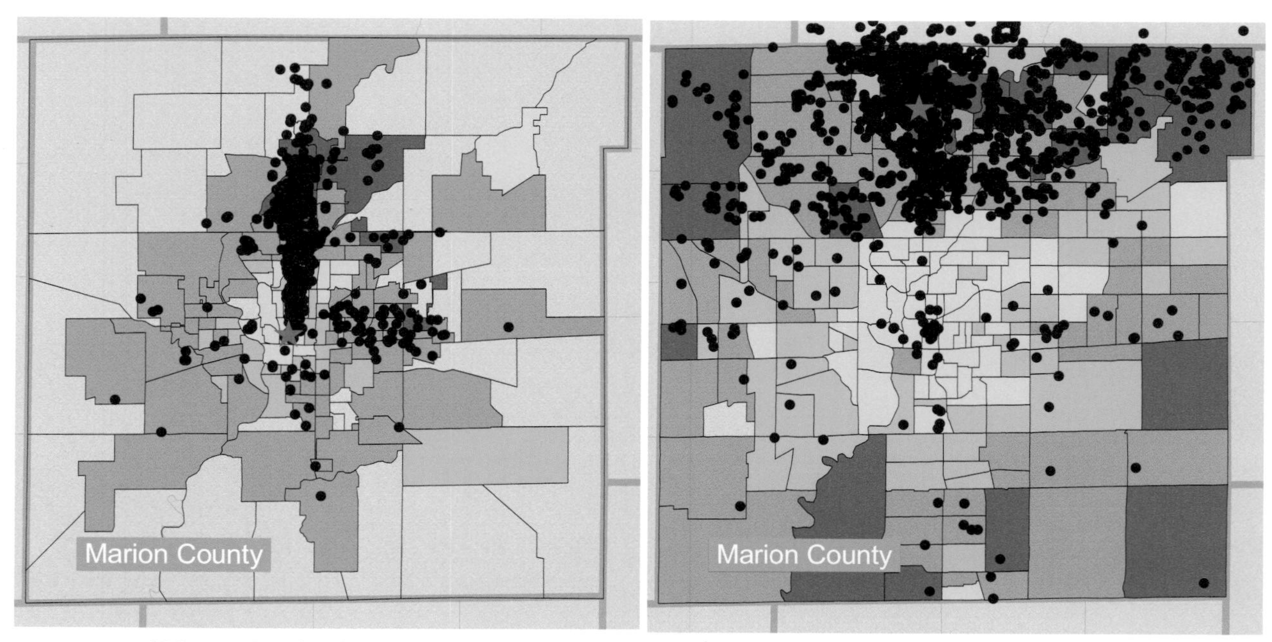

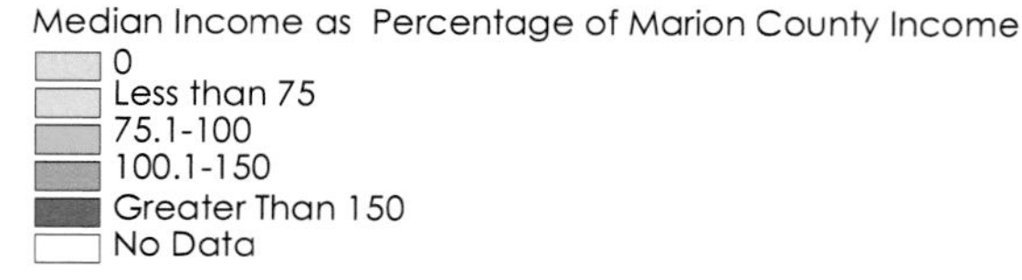

Question: What were the largest religious groups within big cities in 1990?

Step 1: Show largest denomination by county in 1990.
This map shows that Catholics (blue) and Baptists (magenta) dominated most of the Northeast, South, and West.

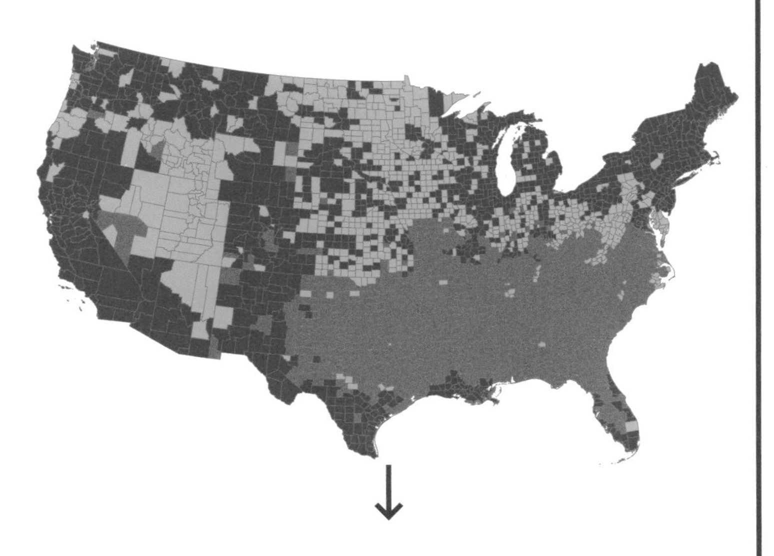

Step 2: Add cities with over 100,000 people to the map.

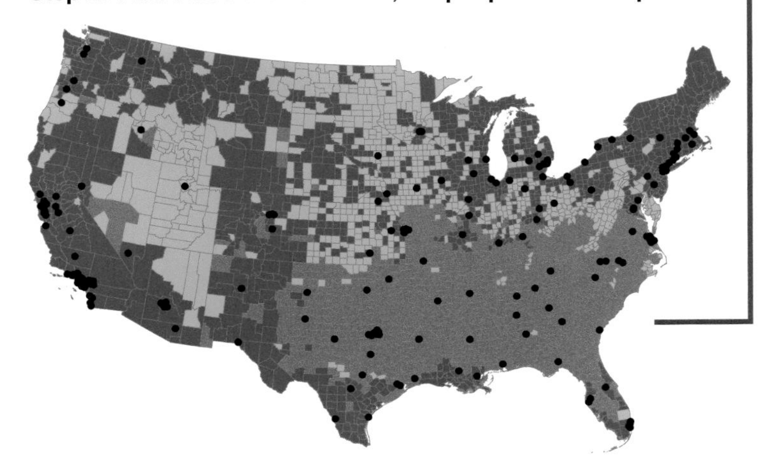

Step 3 (simple query): Select all counties that were either predominantly Baptist or Catholic.
From this selection, you can determine that 75% of counties were either Baptist or Catholic. (42% Baptist, 33% Catholic.)

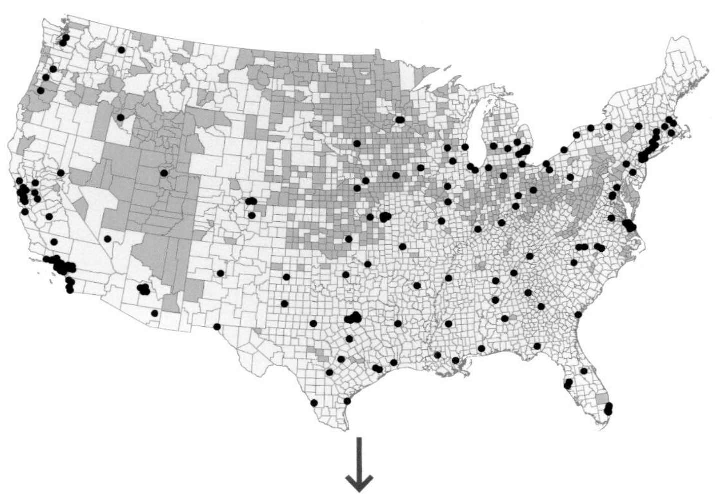

Step 4 (spatial query): Select all counties that were predominantly Baptist or Catholic *and* that contained a large city.
From this selection, you can determine that 95% of counties with large cities were dominated by Baptists or Catholics.

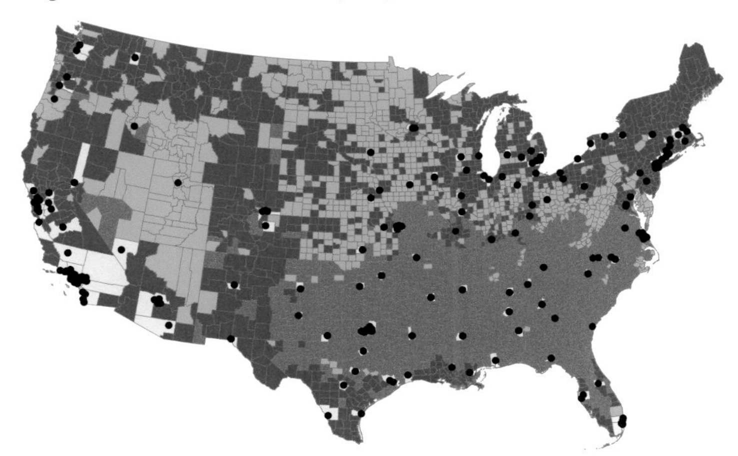

Figure 8. Using GIS to answer spatial questions
This figure walks through the steps one could follow to answer a particular historical question using the North American Religion Atlas. Further spatial queries could ask in which counties Baptists were the largest denomination and Catholics were the second most numerous, or where Baptist membership had increased by more than a certain percentage between two census years.

greatest potential lies in combining religious demographics with other kinds of data to test religion's relationship to secular issues. For example, layering the U.S. Population Census and voting records with the data from the censuses of religion can help explain the relative power of religious beliefs in shaping local and regional politics, the rise and fall of extremist groups such as the Ku Klux Klan, and the persistent influence of some religious groups in particular regions *(figure 9)*.

The process of constructing the North American Religion Atlas raised issues that are likely to occur in any use of GIS for historical research and teaching: the scarcity of digital data, the lack of historical basemaps, the problem of how to handle temporal data, and the difficulties of using qualitative data in a GIS format.

Except for the recent past, most historical information does not exist in digital form. Even when digital records are available, such as the U.S. Census files converted in the 1960s and 1970s by the Inter-University Consortium for Political and Social Research (ICPSR), the resulting databases still may require work to make them suitable for use in a GIS. The North American Religion Atlas team georeferenced the ICPSR files[7] and entered other data manually from carefully marked photocopies of the Census of Religious Bodies.[8] Another ECAI partner, the Centre for Digitisation and Analysis at Queens University in Belfast, reports high accuracy using optical character reader, or OCR, technology.

The absence of accurate historical basemaps in digital form poses a major obstacle to the use of GIS by historians. Not only are reliable maps unavailable for much of the American past, but most maps created before the 1970s exist only on paper. Historians grapple continually with incomplete and imprecise data, but GIS is a technology built on the premise of precise and comprehensive data, especially the spatial location of data. Boundary changes compound the problem for many kinds of social statistics (see chapters 5 and 9).

After examining several digital map series covering the eighteenth through the twentieth century, the North American Religion Atlas selected a 1980 county map file as its framework.[9] This was a cost-effective compromise for the bulk of data planned for the Atlas, but not a solution to the long-term need for high-quality historical GIS maps. The absence of historical basemaps may be the most vexing problem faced by scholars who want to use GIS as a common platform for comparative analysis.

Historians by definition are concerned with issues of place and time. GIS is ideally suited to spatial analysis; it does not accommodate temporal data easily. The problem is not the time marker, which is simply another field in the database, but finding ways to trace and analyze movement across time and space simultaneously. The amount of data required to do such work makes temporal GIS a challenge for all researchers, but the problems are compounded for historians by the imprecision and variability of historical sources. As an interim solution, the North American Religion Atlas provides simultaneous map displays of the same theme at different points in time. TimeMap and other temporal GIS programs aim to facilitate the animation of temporal data, so that researchers can better visualize change over time, as well as enable more complex kinds of spatial analysis of temporally coded data.

The North American Religion Atlas plans to incorporate data from the sixteenth through the nineteenth centuries, data on Mexico and Canada, and georeferenced survey data from the American Religion Data Archive.[10] Tutorials, curricular materials, electronic publications, and development of the user

Figure 9. Predominant U.S. denominations by county, 1971 and 1990
The eye quickly picks out the regional concentration of Catholics in the Northeast and Southwest, the Methodist belt running from central Pennsylvania through Nebraska, and the Lutheran presence in the Upper Midwest. The maps also reveal interesting changes, such as the expansion of Mormon predominance beyond Utah and the growth of Pentecostal and Holiness congregations in the Northwest.

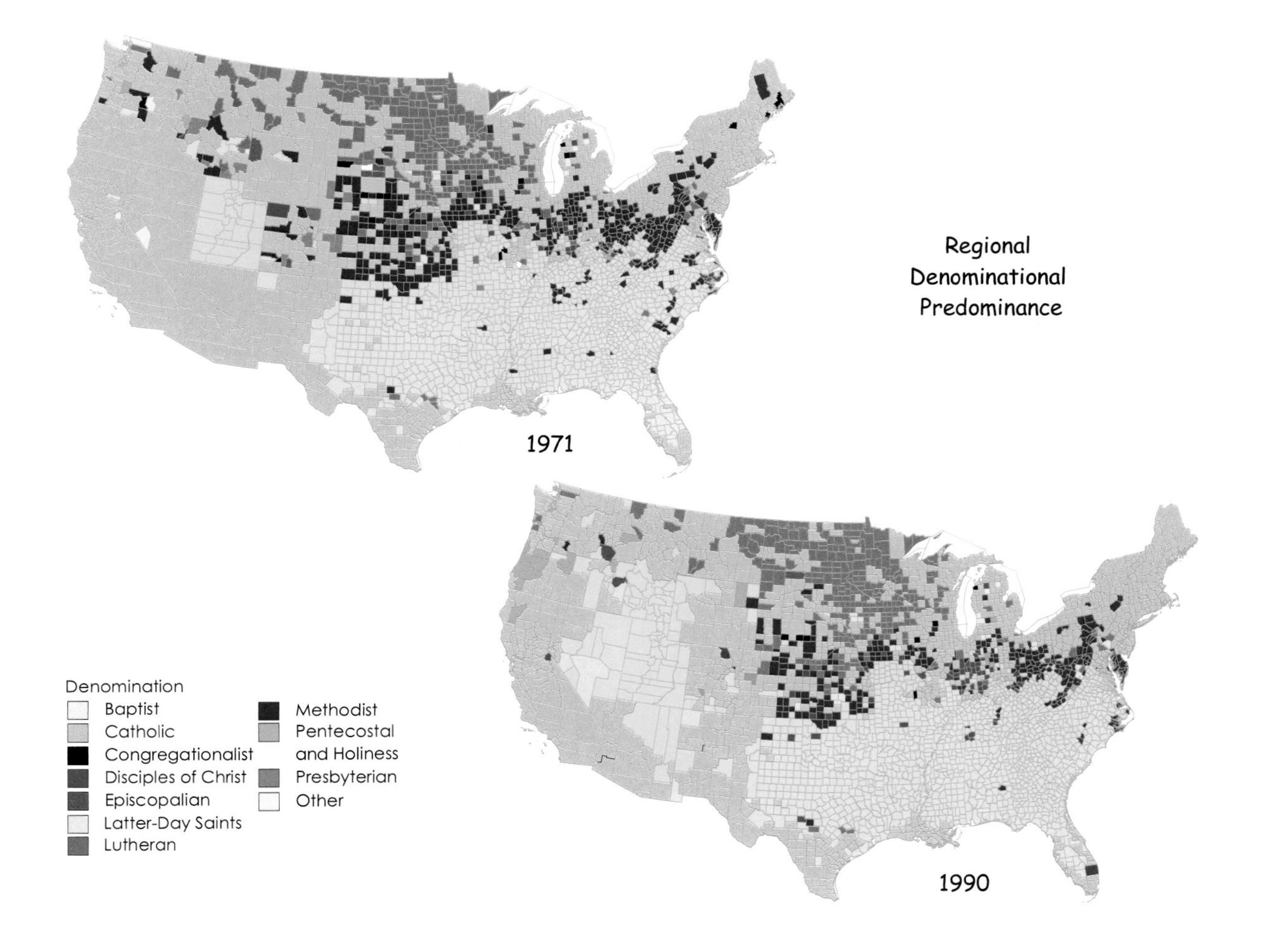

community will make the Atlas a primary resource for teaching as well as research, and a model of ECAI's vision for freely accessible, geographically referenced data for historical scholarship.

In the past, GIS has found users among geographers, city planners, the military, and other groups already thinking spatially. ECAI is now expanding the use of GIS to a new community of scholars who may someday outnumber those now using the technology. GIS will not just mark places on the earth, it will be the software that links library catalogs, data sets, and distributed information. Once the data of humanities and social sciences is being displayed through maps, new kinds of analysis will emerge. We are entering into a world of radical plurality of data. GIS offers us a crucial tool for managing multiplicity.

Acknowledgments

Lewis Lancaster would like to thank Kimberly Carl and Jeanette Zerneke for their assistance in preparing graphics. David Bodenhamer thanks Kevin Mickey for his assistance in preparing this essay, and for their work on the North American Religion Atlas, John Corrigan, Vicki Cummings, Neil Devadasan, Etan Diamond, Dale Drake, Allen Federman, Robert Ferrell, Karen Frederickson, Christian Klaus, Tracy Leavelle, and Ed Stoddard.

Further reading

Fletcher, Roland. *The Limits of Settlement Growth: A Theoretical Outline.* Cambridge: Cambridge University Press, 1995.

Gaustad, Edwin Scott, and Philip L. Barlow. *New Historical Atlas of Religion in America.* New York: Oxford University Press, 2001.

Ghirshman, Roman. *Persian Art, Parthian and Sassanian Dynasties, 249 B.C.–A.D. 651.* Stuart Gilbert and James Emmons, trans. New York: Golden Press, 1962.

Hartwell, Robert. *A Guide to Sources of Chinese Economic History, A.D. 618–1368.* Chicago: Committee on Far Eastern Civilizations, University of Chicago, c. 1964.

Newman, William H., and Peter L. Halvorson. *Atlas of American Religion: The Denominational Era, 1776–1990.* Walnut Creek, Calif.: Alta Mira Press, 2000.

Rothkrug, Lionel. *Religious Practices and Collective Perceptions: Hidden Homologies in the Renaissance and Reformation.* Historical Reflections series, vol. 7, no. 1. Waterloo, Ontario: Dept. of History, University of Waterloo, 1980.

Thomas, Edward. *Early Sassanian Inscriptions, Seals, and Coins.* London: Trubner, 1868.

Notes

1. The ECAI Web site is located at www.ecai.org. ECAI is now a research unit of the International and Area Studies Deanship at the Berkeley campus of the University of California. In 2000, ECAI established a partnership with the California Digital Library, a union library for the nine campuses of the university and the State of California. ECAI also currently collaborates with a number of other libraries, museums, and institutions that use digital technology.

2. Users can access the North American Religion Atlas directly through its home page (www.religionatlas.org) or through links from the Web pages of ECAI or The Polis Center (www.thepoliscenter.iupui.edu).

3. The following variables appear in one or more of the CRB and Glenmary censuses, all available at the county level: adherents by denomination by age (adults and children), gender, race; places of worship and seating capacity; expenditures and debt; Sunday schools and other programs; church property and parsonages; clergy salaries and status (full-time, bivocational); immigrant communications and languages; support of missions and philanthropic institutions; numbers of religiously affiliated hospitals and patients, and schools and pupils.

4. Edwin S. Gaustad, *Historical Atlas of Religion in America* (New York: Harper & Row, 1962); Edwin Scott Gaustad and Philip L. Barlow, *New Historical Atlas of Religion in America* (New York: Oxford University Press, 2001).

5. William H. Newman and Peter L. Halvorson, *Atlas of American Religion: The Denominational Era, 1776–1990* (Walnut Creek, Calif.: Alta Mira Press, 2000).

6. This system, known as Social Assets and Vulnerabilities Indicators (SAVI) database, uses ESRI's Internet Map Server (IMS) software, including MapObjects, a previous-generation Internet map server, and ArcIMS.

7. NARA team members converted ICPSR georeferences to the current FIPS standard to enable mapping FIPS data and to set a standard that would be easily mappable by other researchers.

8. The North American team controlled conversion costs by carefully analyzing the data in the Census of Religious Bodies. It discovered that most of the tables in these large volumes came from four source tables. Converting only those tables and using them to reconstruct the derived tables reduced expenses for this task to a mere fraction of what they would have been otherwise.

9. NARA used the historical GIS boundary files from Louisiana State University in association with the more topologically accurate USCG boundaries to create historical basemaps.

10. Proposed layers include U.S. county boundaries for each census decade, 1790–1890; data from the 1896 U.S. Census of Religious Bodies; data on adherence, budgets, property, and other information for 1600–1900 from the U.S. Census of Religious Bodies; and georeferenced data sets supporting boundary files from the American Religion Data Archives and from denominations represented by the Association for Statisticians of American Religion. The American Religion Data Archive Web site is located at www.TheARDA.com.

Glossary of GIS Terms

For definitions of GIS terms not listed here, see Heather Kennedy, *The ESRI Press Dictionary of GIS Terminology* (ESRI Press, 2001).

A

accuracy The degree to which a value conforms to a specified standard for that value, or the degree to which a measured value is correct. *Compare* precision.

address 1. Also **geocode** A point stored as an x,y location in a geographic data layer, referenced with a unique identifier. 2. [computing] A number that identifies a location in memory where data is stored. 3. A name identifying a site on the Internet or other network.

aerial photograph [remote sensing, photogrammetry] A photograph of the earth's surface taken with a camera mounted in an airplane or balloon. Used in cartography to provide geographical information for basemaps.

altitude [surveying, geodesy] 1. The elevation above a reference datum, usually sea level, of any point on the earth's surface or in the atmosphere. 2. The z-value in a three-dimensional coordinate system.

arc 1. An ordered string of x,y coordinate pairs (vertices) that begin at one location and end at another. Connecting the vertices creates a line. 2. A coverage feature class that represents linear features and polygon boundaries. One line feature can contain many arcs. Arcs are topologically linked to nodes and to polygons. Their attributes are stored in an arc attribute table (AAT). *See also* node.

area 1. Also **polygon** A closed, two-dimensional shape defined by its boundary. 2. The size of a geographic feature measured in square units.

areal interpolation A method of transferring spatial data from one set (source units) to a second set (target units) of overlapping, nonhierarchical areal units. Among other applications, it is used to compare socioeconomic data gathered using different administrative units. In longitudinal studies it can be used to minimize the effects of boundary changes on statistical calculations.

aspect The compass direction that a slope faces, usually measured clockwise from north.

aspect ratio The ratio of the width of an image to its height. A standard computer monitor aspect ratio is 4:3 (rectangular).

atlas [cartography] A collection of maps organized around a theme, such as a world atlas, a national atlas, or a historical atlas.

attribute 1. Information about a geographic feature in a GIS, generally stored in a table and linked to the feature by a unique identifier. Attributes of a river might include its name, length, and average depth. *See* attribute table. 2. Cartographic information that specifies how features are displayed and labeled on a map; the cartographic attributes of the river in (1) above might include line thickness, line length, color, and font.

attribute table A table containing descriptive attributes for a set of geographic features, usually arranged so that each row represents a feature and each column represents one attribute. Each cell in a column stores the value of that column's attribute for that row's feature.

automated cartography Cartography that uses plotters, software, and graphic displays to speed tasks traditionally associated with manual drafting. It does not involve spatial information processing. *Compare* geographic information system.

B

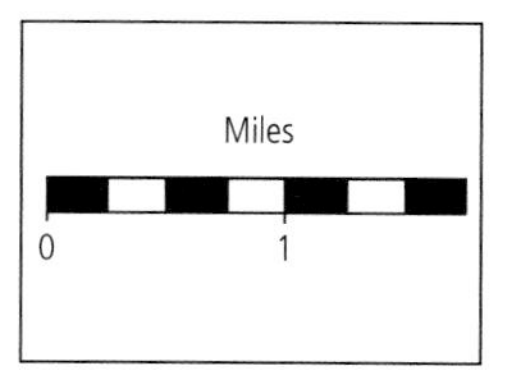

bar scale

bar scale Also **scale bar, graphic scale, linear scale** A scale used to measure distance on a map, marked like a ruler in units proportional to the map's scale.

bathymetry 1. The science of measuring and charting the depths of water bodies. 2. The measurements so obtained.

bit map Also **bit image** An image format in which each pixel on the screen is represented by one or more bits. The number of bits per pixel determines the shades of gray or number of colors that a bit map can represent.

buffer 1. [topology] A polygon enclosing a point, line, or polygon at a specified distance. 2. [computing] A storage area, usually in RAM, that holds data while it is transferred from one location to another.

C

cadastral survey A boundary survey taken for the purposes of taxation.

cadastre A public record of the dimensions and value of land parcels, used to record ownership and calculate taxes.

Cartesian coordinate system [geometry] A system of reference in which location is measured along the planes created by two or three mutually perpendicular intersecting axes. In two dimensions, points are described by their positions in relation to two axes, x and y. A third axis, z, is added to measure locations in three dimensions. Relative measure of distance, area, and direction are constant throughout the system. Named after René Descartes, who originated the two-dimensional system in the seventeenth century.

cartogram A diagram or abstract map in which geographical areas are exaggerated or distorted in proportion to the value of an attribute.

cartographic elements The primitive components that make up a map, such as the neatline, legend, scale, titles, and figures.

cartography The design, compilation, drafting, and reproduction of maps.

cell 1. The smallest square in a grid. Each cell usually has an attribute value associated with it. 2. A pixel.

cell size Also **pixel size** The area on the ground covered by a single pixel in an image, measured in map units.

census block [demography] The smallest geographic unit used by the U.S. Census Bureau for reporting census data and for generating geographic base files such as DIME and TIGER files. A block is enclosed by any natural or human-made features that form a logical boundary, such as roads, political boundaries, or shorelines.

census tract A geographical area that combines adjacent census blocks into a group of approximately four thousand people.

choropleth A thematic map in which areas are colored or shaded to reflect the density of the mapped phenomenon or to symbolize classes within it.

class 1. A group or category of attribute values. 2. Pixels in a raster file that represent the same condition.

column Also **field, item** The vertical dimension of a table. Each column stores the values of one type of attribute for all of the records, or rows, in the table. All of the values in a given column are of the same data type (e.g., number, string, blob, date). *See* attribute table.

command An instruction to a computer program, usually one word or concatenated words or letters, issued by the user from a control device such as a keyboard or read from a file by a command interpreter. Menu items on a GUI are also often referred to as commands.

continuous data Data such as surface elevation or temperature that varies without discrete steps. Since computers store data discretely, continuous data is usually represented by TINs, rasters, or contour lines, so that any location has either a specified value or one that can be derived.

continuous tone image A photograph that has not been screened and so displays all the tones from black to white or dark to light color. See also halftone image, dot screen.

contour interval The difference in elevation between two contour lines.

contour line A line drawn on a map connecting points of equal elevation above a datum, usually mean sea level.

control points 1. Also **ground control points** Points on a map representing locations whose coordinates are known in some system of ground measurement such as latitude and longitude. 2. Points in a data layer representing known locations, used to register map sheets for digitizing and to transform digitizer coordinates to a common coordinate system.

coordinates 1. The x- and y-values that define a location in a planar coordinate system. 2. The x-, y-, and z-values that define a location in a three-dimensional coordinate system.

coordinate system A reference system consisting of a set of points, lines, and/or surfaces, and a set of rules, used to define the positions of points in space in either two or three dimensions. *See also* geographic coordinate system, planar coordinate system.

coverage An ArcInfo vector data storage format. A coverage stores the location, shape, and attributes of geographic features, and usually represents a single theme such as soils, streams, roads, or land use. Map features are stored as both primary features (e.g., arcs, polygons, and points) and secondary features (e.g., tics, links, and annotation). The attributes of geographic features are stored independently in feature attribute tables.

cross tabulation Comparing attributes in different map layers according to location.

cultural features Human-made features, on a map or on the ground.

cultural geography Geography that studies human culture and its effects on the earth.

D

data Any collection of related facts arranged in a particular format; often, the basic elements of information that are produced, stored, or processed by a computer.

data automation Any electronic, electromechanical, or mechanical means for recording, communicating, or processing data.

database One or more structured sets of persistent data, managed and stored as a unit and generally associated with software to update and query the data. A simple database might be a single file with many records, each of which references the same set of fields. A GIS database includes data about the spatial locations and shapes of geographic features recorded as points, lines, areas, pixels, grid cells, or TINs, as well as their attributes.

data capture Any operation that converts digital or analog data into computer-readable form. Geographic data can be downloaded directly into a GIS from sources such as remote sensing or GPS, or it can be digitized, scanned, or keyed in manually from paper maps or photographs.

data conversion Translating data from one format to another, usually in order to move it from one system to another.

data entry The transfer of data into a computer by manual key entry.

data integration Combining databases or data files from organizations that collect information about the same entities (such as properties, census tracts, or sewer lines). Doing so prevents redundant work and creates new ways to analyze the information.

data set Any collection of data with a common theme.

data type 1. In a database table, the types of data that columns and variables can store. Examples include character, floating point, and integer. 2. [programming] Specifications of the possible range of values of a data set, the operations that can be performed on it, and the way the values are stored in memory.

decimal degrees Degrees of latitude and longitude expressed in decimals instead of in degrees, minutes, and seconds. Decimal degrees are computed with the formula

$$\textit{decimal degrees} = \textit{degrees} + \textit{minutes}/60 + \textit{seconds}/3{,}600$$

Using this formula, 73° 59′ 15″ longitude is equal to 73.9875 decimal degrees.

degree A unit of angular measure, represented by the symbol °. The circumference of a circle contains 360 degrees.

degrees/minutes/seconds (DMS) A measurement of degrees of latitude and longitude in which each degree is divided into sixty minutes and each minute is divided into sixty seconds.

DEM *See* digital elevation model.

demographics The statistical characteristics (such as age, birth rate, and income) of a human population.

demography The study of human vital and social statistics, such as births, deaths, health, marriage, and welfare.

desktop GIS Mapping software that runs on a personal computer and can display, query, update, and analyze geographic locations and the information linked to those locations.

digital [computing] Also, often, **binary** Data processed in discreet, quantified units. Most computers process information as combinations of binary digits, or bits.

digital elevation model (DEM) Also **digital terrain model (DTM)** 1. The representation of continuous elevation values over a topographic surface by a regular array of z-values, referenced to a common datum. Typically used to represent terrain relief. 2. The database for elevation data by map sheet from the National Mapping Division of the U.S. Geological Survey.

digital image [remote sensing, photogrammetry, graphics] An image stored in binary form and divided into a matrix of pixels, each of which consists of one or more bits of information that represent either the brightness, or the brightness and color, of the image at that point.

digital line graph (DLG) Vector data files of transportation, hydrography, contour, and public land survey boundaries from USGS basemaps.

digital terrain model (DTM) *See* digital elevation model.

digitize To convert the shapes of geographic features from media such as paper maps or raster imagery into vector x,y coordinates. *See* digitizer.

digitizer 1. (Manual) A device consisting of a tablet and a handheld cursor that converts electronic signals from positions on the tablet surface to digital x,y coordinates, yielding vector data consisting of points, lines, and polygons. 2. The title of the person who uses a digitizer. 3. (Video) An optical device that translates an analog image into an array of digital pixel values. A video digitizer can be used in place of a manual digitizer, but since it produces a raster image, additional software must be used to convert the data into vector format before topological analysis can be done.

directory An area of a computer disk that holds a set of data files and/or other directories. Directories are arranged in a tree structure, in which each branch is a subdirectory of its parent branch. The location of a directory is specified with a pathname, for example C:\gisprojects\shrinkinglemurhabitat\grids.

display resolution The number of pixels displayed on a monitor, measured horizontally and vertically (for example, 1,024 by 768).

distance The amount of space between two things that may or may not be connected, such as two points. Differentiated from length, which always implies a physical connection.

distortion On a map or an image, the misrepresentation of shape, area, distance, or direction of or between geographic features when compared to their true measurements on the curved surface of the earth.

distribution 1. The amount or frequency of the occurrence of a thing or things within a given area. 2. The set of probabilities that a variable will have a particular value.

DLG *See* digital line graph.

dot distribution map A map that uses dots or other symbols to represent the presence, quantity, or value of a thing in a specific area. Symbols whose sizes differ in relation to the phenomenon being mapped are called proportional symbols.

drape A perspective or panoramic rendering of a two-dimensional image superimposed on a three-dimensional surface.

E

elevation Also **altitude** The vertical distance of a point or object above or below a reference surface or datum (generally mean sea level).

export 1. To move data from one computer system to another, and often, in the process, from one file format to another. 2. An ArcInfo command that creates an interchange file, or E00 file, for transferring coverages between different systems.

F

feature 1. An object in a landscape or on a map. 2. A shape in a spatial data layer, such as a point, line, or polygon, that represents a geographic object.

field 1. Also **item** A vertical column in a table that represents some characteristic for all of the records in the table, given in numbers or words. 2. The place in a database record, or in a graphical user interface, where data can be entered. 3. A synonym for surface.

file Information stored on disk or tape. A file may be a collection of data, a document (text file), or a program (executable file). It generally resides within a directory, and always has a unique name.

friction surface A raster surface whose pixel values are calculated by algorithms that model things like slope, distance, bodies of water, soil type, and elevation, that can impede movement by people, objects, or fluids.

G

gazetteer 1. A list of geographic places and their coordinates, along with other information such as area, population, and cultural statistics. 2. A geographical search engine that enables users to search for information by place.

generalization 1. Reducing the number of points in a line without losing its essential shape. 2. Enlarging and resampling cells in a raster format. 3. [cartography] Any reduction of information so that a map is clear and uncluttered when its scale is reduced.

geocode A code representing the location of an object, such as an address, a census tract, a postal code, or x,y coordinates.

geodesy The science that determines the size and shape of the earth and measures its gravitational and magnetic fields.

geodetic survey A survey that takes the figure and size of the earth into account, used to precisely locate horizontal and vertical positions suitable for controlling other surveys.

geographic coordinates Locations on the surface of the earth expressed in degrees of latitude and longitude.

geographic coordinate system [geodesy, navigation, surveying] A reference system using latitude and longitude to define the locations of points on the surface of a sphere or spheroid.

geographic information Qualitative or quantitative information that can be located on the surface of the earth.

geographic information system (GIS) A collection of computer hardware, software, and geographic data for capturing, storing, updating, manipulating, analyzing, and displaying all forms of geographically referenced information.

geography 1. The study of the earth's surface, especially how climate and elevation interact with soil, vegetation, and animal populations. 2. The geographic features of an area. 3. A word game in which each player in rotation says aloud a geographic place-name beginning with the last letter of the place-name mentioned by the preceding player.

georeference To assign coordinates from a known reference system, such as latitude/longitude, UTM, or State Plane, to the page coordinates of an image or a planar map. *See also* rectification.

GIS *See* geographic information system.

Global Positioning System (GPS) A constellation of twenty-four satellites, developed by the U.S. Department of Defense, that orbit the earth at an altitude of 20,200 kilometers. These satellites transmit signals that allow a GPS receiver anywhere on earth to calculate its own location. The Global Positioning System is used in navigation, mapping, surveying, and other applications where precise positioning is necessary.

GPS *See* Global Positioning System.

graduated color The use of a range of colors to indicate a progression of numeric values. For example, differences in population density could be represented by increasing the saturation of a single color, and temperature changes could be represented by colors ranging from blue to red.

graduated symbol A set of symbols sized to correspond to the amount of the attribute they represent. For example, larger rivers could be represented by thicker lines, and denser populations by larger dots.

graphic elements [cartography] The basic characteristics of any map symbol: size, position, shape, spacing, hue, value, saturation, brightness, orientation, and pattern.

graticule 1. [mapping, geodesy] A network of longitude and latitude lines on a map or chart that relates points on a map to their true locations on the earth. 2. [astronomy] A glass plate or cell with a grid or cross wires on it that rests in the focal plane of the eyepiece of a telescope, used to locate and measure celestial objects.

grid 1. Equally sized square cells arranged in rows and columns. Each cell contains a value for the feature it covers. *See also* raster. 2. [cartography] Any network of parallel and perpendicular lines superimposed on a map, usually named after the map's projection, such as a Lambert grid, or transverse Mercator grid.

grid cell 1. A single square in a grid that represents a portion of the earth, such as a square meter or square mile. Each grid cell has a value for the feature or attribute that it covers, such as soil type, census tract, or vegetation class. 2. A pixel.

grid reference system A reference system that uses a rectangular grid to assign x,y coordinates to individual locations. *See* Cartesian coordinate system.

ground control Also **control mapping** [surveying, remote sensing, photogrammetry] A system of points with established positions, elevations, or both, used as fixed references in relating map features, aerial photographs, or remotely sensed images.

ground truth 1. Ground control. 2. Testing or verifying archival information or other sources in the field.

H

hachures [cartography] Lines on a map that indicate the direction and steepness of slopes. For steep slopes the lines are short and close together; for gentle slopes they are longer, lighter, and farther apart. Contours, shading, and hypsometric tints have largely replaced hachuring on modern maps.

halftone image A continuous tone image photographed through a fine screen that converts it into uniformly spaced dots of varying size while maintaining all the gradations of highlight and shadow. The size of the dots varies in proportion to the intensity of the light passing through them.

hillshading *See* relief shading.

histogram A graph showing the distribution of values in a set of data. Individual values are displayed along a horizontal axis, and the frequency of their occurrence is displayed along a vertical axis.

hot link A link that connects a geographic feature to an external image, text, or executable file. When the feature is clicked, the file runs or is displayed on-screen.

hydrographic data 1. Information about depths, depth contours, and elevations of oceanic features. 2. Data layer with information about rivers, streams, lakes, and other bodies of water.

hypsometric map A map showing relief, whether by contours, hachures, shading, or tinting.

I

import To load data from one computer system or application into another. Importing often involves some form of data conversion.

index A data structure used to speed the search for records in a database or for spatial features in geographic data sets. In general, unique identifiers stored in a key field point to records or files holding more detailed information.

information system A system that contains or is related to a database of information and also provides the means of data storage, retrieval, and analysis, so that a user may query and receive answers from the database.

inset map A small map set within a larger map. An inset map might show an area that does not fit neatly into the main map, or a detail of part of the map at a larger scale, or the context of the area covered by the map at a smaller scale.

interface For the purpose of data communication, a hardware and software link that connects two computer systems, two applications, a computer and its peripherals, or a computer and its user. *See* graphical user interface, command-line interface.

integration *See* data integration.

isometric line [cartography] An isarithm drawn according to known values, either sampled or derived, that can occur at points. Examples of sampled quantities that can occur at points are elevation above sea level, an actual temperature, or an actual depth of precipitation. Examples of derived values that can occur at points are the average of temperature over time for one point or the ratio of smoggy days to clear days for one point.

iterative mapping Producing many maps as answers to slightly varied queries of a GIS database.

J

join Appending the fields of one table to those of another through a common item. A join is usually used to attach more attributes to the attribute table of a geographic layer. *See also* relational join, spatial join.

L

label Text placed next to a feature on a map to describe or identify it.

landform Any natural feature of the land having a characteristic shape, including major forms such as plains and mountains and minor forms such as hills and valleys.

Landsat [remote sensing] Earth-orbiting satellites developed by NASA that gather imagery for land-use inventory, geological and mineralogical exploration, crop and forestry assessment, and cartography.

land use The classification of land according to how it is used; for example, agricultural, industrial, residential, urban, rural, or commercial. Natural features of the land such as forest, pastureland, brushland, and bodies of water are also often classified in this manner.

large scale [cartography] Generally, a map scale whose representative fraction is 1:50,000 or larger. A large-scale map shows a small area on the ground at a high level of detail. *See also* scale, small scale, medium scale, representative fraction.

latitude [navigation, geodesy] The angular distance along a meridian north or south of the equator, usually measured in degrees. Lines of latitude are also called parallels.

latitude–longitude Also **lat/long, lat/lon** [navigation, geodesy] The most commonly used spherical reference system for locating positions on the earth. Latitude and longitude are angles measured from the equator and the prime meridian to locations on the earth's surface. Latitude measures angles in a north–south direction; longitude measures angles in the east–west direction.

layer 1. A set of vector data organized by subject matter, such as roads, rivers, or political boundaries. Vector layers act as digital transparencies that can be laid atop one another for viewing or spatial analysis. 2. A set of raster data representing a particular geographic area, such as an aerial photograph or a remotely sensed image. In both (1) and (2), layers covering the same geographical space are registered to one another by means of a common coordinate system. 3. A file that stores symbology and display information for a given vector or raster data set. The layer does not actually contain the data, but points to its physical location.

layout [cartography] 1. The way map elements such as the title, legend, and scale bar are arranged on a printed map. 2. An on-screen document where said map elements are arranged for printing.

legend [cartography] The reference area on a map that lists and explains the colors, symbols, line patterns, shadings, and annotation used on the map, and often includes the map's scale, origin, and projection.

line Also **linear feature** A shape having length and direction but no area, connecting at least two x,y coordinates. Lines represent geographic features too narrow to be displayed as an area at a given scale, such as contours, street centerlines, or streams, or linear features with no area, such as state and county boundary lines.

location Also **position** A point on the earth's surface or in geographical space described by x-, y-, and z-coordinates, or by other precise information such as a street address.

longitude The angular distance, expressed in degrees, minutes, and seconds, of a point on the earth's surface east or west of a prime meridian (usually the Greenwich meridian). All lines of longitude are great circles that intersect the equator and pass through the north and south poles.

longitudinal study A study of change over time, typically following a given population over a series of years or decades or examining change within a given area.

M

map 1. A graphic depiction on a flat surface of the physical features of the whole or a part of the earth or other body, or of the heavens, using shapes or photographic imagery to represent objects, and symbols to describe their nature; at a scale whose representative fraction is less than 1:1, generally using a specified projection and indicating the direction of orientation. 2. Any graphical presentation of geographic or spatial information.

map generalization *See* generalization.

map library A collection of geographic data partitioned spatially as a set of tiles and thematically as a set of layers, indexed by location for rapid access.

map projection [cartography] A mathematical model that transforms the locations of features on the earth's curved surface to locations on a two-dimensional surface. It can be visualized as a transparent globe with a lightbulb at its center casting lines of latitude and longitude onto a sheet of paper. Generally, the paper is either flat and placed tangent to the globe (a planar or azimuthal projection), or formed into a cone or cylinder and placed over the globe (cylindrical and conical projections). Every map projection distorts distance, area, shape, direction, or some combination thereof.

map query *See* query.

medium scale Generally, a map scale whose representative fraction is between 1:50,000 and 1:500,000. *See* scale, large scale, small scale, representative fraction.

meridian [navigation, geodesy] A great circle on the earth that passes through the poles, often used synonymously with longitude. From a prime meridian or 0 degrees longitude (usually the meridian that runs through the Royal Observatory in Greenwich, England), measures of longitude are negative to the west and positive to the east, where they meet halfway around the globe at the line of 180 degrees longitude.

metadata Information about a data set. Metadata for geographical data may include the source of the data; its creation date and format; its projection, scale, resolution, and accuracy; and its reliability with regard to some standard.

metes and bounds [surveying] The limits of a land parcel identified as relative distances and bearings from natural or human-made landmarks. Metes and bounds surveying is often used for areas that are irregularly shaped.

minute Also **angular minute, minute of arc** An angle equal to one sixtieth of a degree of latitude or longitude and containing sixty seconds.

mosaic 1. Maps of adjacent areas with the same projection, datum, ellipsoid, and scale whose boundaries have been matched and dissolved. 2. An image made by assembling individual images or photographs of adjacent areas.

node 1. The beginning and ending points of an arc, topologically linked to all the arcs that meet there. 2. In graph theory, the location at which three or more lines connect. 3. One of the three corner points of a triangle in a TIN, topologically linked to all triangles that meet there. Each sample point in a TIN becomes a node in the triangu-lation. 4. [computing] The point at which a computer, or other addressable device, attaches to a communications network.

O

orthophotograph Also **digital orthophoto** [remote sensing, photogrammetry] A perspective aerial photograph from which distortions owing to camera tilt and ground relief have been removed. An orthophotograph has the same scale throughout and can be used as a map.

overlay Superimposing two or more maps registered to a common coordinate system, either digitally or on a transparent material, in order to show the relationships between features that occupy the same geographic space.

P

pan To move an on-screen display window up, down, or sideways over a map or image without changing the viewing scale.

photogrammetry Recording, measuring, and plotting electromagnetic radiation data from aerial photographs and remote sensing systems against land features identified in ground control surveys, generally in order to produce planimetric, topographic, and contour maps.

physical geography The study of the natural features of the earth's surface.

pixel (picture element) 1. [computing] The smallest addressable hardware unit on a display device. 2. The smallest unit of information in an image or raster map. Usually rectangular, pixel is often used synonymously with cell.

pixel coordinate system An image coordinate system whose measurement units are pixels. In contrast to most map coordinate systems, the origin (0,0) usually lies in the upper left corner of the image and the y-values increase as they go down the page. *Compare* Cartesian coordinate system, planar coordinate system.

planar coordinate system A two-dimensional coordinate system that locates features according to their distance from an origin (0,0) along two axes, a horizontal axis (x) representing east–west and a vertical axis (y) representing north–south.

plane survey A survey of a small area that does not take the curvature of the earth's surface into account.

planimetric 1. Two-dimensional; showing no relief. 2. A map that gives only the x,y locations of features and represents only horizontal distances correctly. *Compare* topographic.

plat A survey diagram, drawn to scale, of the legal boundaries and divisions of a tract of land.

plotter A device that draws an image onto paper or transparencies, either with colored pens or by drawing an image of electrostatically charged dots and fusing it onto the paper with toner. A flatbed plotter holds the paper still and draws along its x- and y-axes, a drum plotter draws along one axis and rolls the paper over a cylinder along the other axis, and a pinch roller draws along one axis and moves the paper back and forth on the other axis over small rollers.

point Also **point feature** A single x,y coordinate that represents a geographic feature too small to be displayed as a line or area at that scale.

polygon A two-dimensional closed figure with at least three sides that represents an area. *See also* area.

position Also **location** The latitude, longitude, and altitude (x,y,z) of a point, often accompanied by an estimate of error. It may also refer to an object's orientation (facing east, for example) without referring to its location.

precision 1. The number of significant digits used to store numbers, particularly coordinates. *See* single and double precision. 2. The exactness with or detail in which a value is expressed, right or wrong. *Compare* accuracy. 3. A statistical measure of repeatability, usually expressed as the variance of repeated measures about the mean.

projected coordinates Latitude and longitude coordinates projected to x,y coordinates in a planar coordinate system. *Compare* geographic coordinates.

projection *See* map projection.

Q

quadrangle (quad) Also **topographic map, topo** A rectangular map bounded by lines of latitude and longitude, often a map sheet in either the 7.5-minute or 15-minute series published by the U.S. Geological Survey.

qualitative 1. Data grouped by kind, not by amount or rank, such as soil by type or animals by species. 2. A map that shows only how data is distributed spatially. A dot map of all cities in the United States with no regard to size or population would be a qualitative map.

quantitative 1. Data that can be measured, such as air temperature or wheat production. 2. A map showing the spatial distribution of measurable data, such as a map of counties shaded by population.

query Also **attribute query, map query, spatial query, logical query** A query (or attribute query) is a statement or logical expression used to select features or records from a database. A map query asks spatial or logical questions of the data in a GIS. A spatial query selects features on the basis of their location or spatial relationship to each other. A logical query selects features whose attributes meet specific criteria; for example, all polygons whose value for area is greater than 10,000, or all arcs whose value for name is "Main St."

R

raster 1. A spatial data model made of rows and columns of cells. Each cell contains an attribute value and location coordinates; the coordinates are contained in the ordering of the matrix, unlike a vector structure which stores coordinates explicitly. Groups of cells that share the same value represent geographic features. *See also* grid; *compare* vector. 2. The illumination on a video display produced by repeatedly sweeping a beam of electrons over the phosphorescent screen line by line from top to bottom. 3. Also **raster image, bit-map image, image** A matrix of pixels whose values represent the level of energy reflected or emitted by the surface being photographed, scanned, or otherwise sensed.

rasterization Also **vector-to-raster conversion** The conversion of points, lines, and polygons into cell data.

record 1. A row in a database or in an attribute table that contains all of the attribute values for a single entity. 2. [computing] Also **line** An ordered set of fields in a file.

rectification [georeferencing] 1. Referencing features in an image or grid to a geographic coordinate system. 2. Converting an image or map from one coordinate system to another. 3. Removing the effects of tilt or relief from a map or image. *See also* georeference.

region A coverage feature class that can represent a single area feature as more than one polygon.

register 1. To align two or more maps or images so that equivalent geographic coordinates coincide. 2. To link map coordinates to ground control points.

relational database Data stored in tables that are associated by shared attributes. Any data element can be found in the database through the name of the table, the attribute (column) name, and the value of the primary key. In contrast to hierarchical and network database structures, the data can be arranged in different combinations.

relief Elevations and depressions of the earth's surface, including those of the ocean floor. Relief can be represented on maps by contours, shading, hypsometric tints, digital terrain modeling, or spot elevations.

relief map A map that is or appears to be three-dimensional.

relief shading Also **hillshading** 1. [cartography] Shadows drawn on a map to simulate the effect of the sun's rays over the land. 2. On a grid, the same effect achieved by assigning an illumination value from 0 to 255 to each cell according to a specified azimuth and altitude for the sun.

remote sensing Collecting and interpreting information about the environment and the surface of the earth from a distance, primarily by sensing radiation that is naturally emitted or reflected by the earth's surface or from the atmosphere, or by sensing signals transmitted from a satellite and reflected back to it. Examples of remote sensing methods include aerial photography, radar, and satellite imaging.

remote sensing imagery Imagery acquired from satellites and aircraft. Examples include panchromatic, infrared black-and-white, and infrared color photographs, and thermal infrared, radar, and microwave imagery.

representative fraction The ratio of a distance on a map to the equivalent distance measured in the same units on the ground. A scale of 1:50,000 means that one inch on the map equals 50,000 inches on the ground. *See also* scale.

resolution 1. The area represented by each pixel in an image. 2. The smallest spacing between two display elements, expressed as dots per inch, pixels per line, or lines per millimeter. 3. The detail with which a map depicts the location and shape of geographic features. The larger the map scale, the higher the possible resolution. As scale decreases, resolution diminishes and feature boundaries must be smoothed, simplified, or not shown at all; for example, small areas may have to be represented as points.

row 1. A horizontal record in an attribute table. 2. A horizontal group of cells in a grid, or pixels in an image.

rubber sheeting Also **warping, elastic transformation** Mathematically stretching or shrinking a portion of a map or image in order to align its coordinates with known control points.

S

satellite imagery *See* remote sensing imagery.

scale The ratio or relationship between a distance or area on a map and the corresponding distance or area on the ground. *See* bar scale, verbal scale, representative fraction.

scale factor 1. The ratio of the actual scale at a particular place on a map to the stated scale of the map. 2. A value, usually less than one, that converts a tangent projection to a secant projection.

scanner 1. A device that sweeps a light beam across the surface of a map or image and records the information in raster format. 2. A device that records the radiation reflected or emitted by the earth's surface and processes it as per (1).

shaded relief image A raster image that shows light and shadow on terrain from a given angle of the sun.

shading Graphic patterns such as cross hatching, lines, or color or grayscale tones that distinguish one area from another on a map.

shapefile A vector file format for storing the location, shape, and attributes of geographic features. It is stored in a set of related files and contains one feature class.

small scale Generally, a map scale whose representative fraction is 1:500,000 or smaller. A small-scale map shows a relatively large area on the ground with a low level of detail. *See* scale, large scale, medium scale, representative fraction.

spatial analysis Studying the locations and shapes of geographic features and the relationships between them. It traditionally includes overlay and contiguity analysis, surface analysis, linear analysis, and raster analysis.

spatial autocorrelation A statistical measure of the spatial law that everything is related to everything else, but that things near each other are more related than those far apart. If features close to each other have similar attributes, the pattern is said to show positive spatial autocorrelation. If features that are near one another have fewer similarities than features that are farther apart, the pattern shows negative spatial autocorrelation.

spatial data 1. Information about the locations and shapes of geographic features, and the relationships between them; usually stored as coordinates and topology. 2. Any data that can be mapped.

spatial modeling Any procedures that use the spatial relationships between geographic features to simulate real-world conditions, such as geometric modeling (generating buffers, calculating areas and perimeters, and calculating distances between features), coincidence modeling (topological overlay), and adjacency modeling (pathfinding, redistricting, and allocation).

statistical surface Ordinal, interval, or ratio data represented as a surface. The height of each area is proportional to a numerical value.

surface A geographic phenomenon represented as a set of continuous data, such as elevation or air temperature. Models of surfaces can be built from sample points, isolines, bathymetry, and the like.

surveying Measuring physical, chemical, or geometric characteristics of the earth. Surveys are often classified by the type of data studied or by the instruments or methods used. Examples include geodetic, geologic, topographic, hydrographic, land, geophysical, soil, mine, and engineering surveys.

symbol [cartography] A mark used to represent a geographic feature on a map. Symbols can look like what they represent (tiny trees, railroads, houses) or they can be abstract shapes (points, lines, polygons). They are usually explained in a map legend.

T

thematic data Features of one type that are generally placed together in a single geographical layer. *See* theme.

theme 1. A vector layer of related geographic features, such as streets, rivers, or parcels, that when juxtaposed with other themes can be used in overlay analysis. 2. A raster layer of geographic information, such as an image or a grid.

Thiessen polygons Also **Voronoi diagrams, Dirichlet tessellations** Polygons generated from a set of points, defined by the perpendicular bisectors of the lines between all points and drawn so that each polygon bounds the region that is closer to one point than to any adjacent point.

TIGER (Topologically Integrated Geographic Encoding and Referencing) The nationwide digital database developed for the 1990 census, succeeding the DIME format. TIGER files contain street address ranges, census tracts, and block boundaries.

TIN *See* triangulated irregular network.

topographic 1. Having elevation. 2. A map showing relief, often as contour lines, along with other natural and human-made features. 3. Map sheets published by the U.S. Geological Survey in the 7.5-minute or 15-minute quadrangle series.

topography The shape or configuration of the land, represented on a map by contour lines, hypsometric tints, and relief shading.

topology 1. The spatial relationships between connecting or adjacent features in a geographic data layer. Topological relationships are used for spatial modeling operations that do not require coordinate information. 2. [geometry, mathematics] The branch of geometry that deals with the properties of a figure that remain unchanged even when the figure is bent, stretched, or otherwise distorted.

triangulated irregular network (TIN) A vector data structure that partitions geographic space into contiguous, nonoverlapping triangles. The vertices of each triangle are data points with x-, y-, and z-values; elevation values at these points are interpolated to create a continuous surface.

U

United States Geological Survey (USGS) The national mapping agency of the United States that produces paper maps, digital maps, and DEMs at a variety of scales, including 1:24,000, 1:100,000, 1:250,000, and 1:1 million. Its national map database consists of 1:100,000 maps, available as digital line graph (DLG) and TIGER files.

V

vector 1. A data structure used to represent linear geographic features. Features are made of ordered lists of x,y coordinates and represented by points, lines, or polygons; points connect to become lines, and lines connect to become polygons. Attributes are associated with each feature (as opposed to a raster data structure, which associates attributes with grid cells). *Compare* raster. 2. Any quantity that has both magnitude and direction.

vertex 1. One of a set of ordered x,y coordinates that constitutes a line. 2. The junction of lines that form an angle. 3. The highest point of a feature.

vertical exaggeration A multiplier applied uniformly to the z-values in a three-dimensional model to enhance or minimize the natural variations of its surface. Vertical exaggeration is generally applied more to flat regions than to mountainous ones.

viewshed analysis A method of spatial analysis that calculates and displays in map form which areas are visible and which are not from a specified x,y,z position.

voxel A three-dimensional pixel; bulky to store.

X

x-axis 1. In a planar coordinate system, the horizontal line that runs to the right and left (east–west) of the origin (0,0). Numbers to the east of the origin are positive and numbers to the west are negative. 2. In a spherical coordinate system, the x-axis is in the equatorial plane and passes through 0 degrees longitude. *See* y-axis, z-axis, Cartesian coordinate system. 3. On a chart, the horizontal axis.

x,y coordinates A pair of numbers expressing a point's horizontal and vertical distance along two orthogonal axes, from the origin (0,0) where the axes cross. Usually, the x-coordinate is measured along the east–west axis and the y-coordinate is measured along the north–south axis.

x,y,z coordinates In a planar coordinate system, three coordinates that locate a point by its distance from an origin (0,0,0) where three orthogonal axes cross. Usually, the x-coordinate is measured along the east–west axis, the y-coordinate is measured along the north–south axis, and the z-coordinate measures height or elevation.

Y

y-axis 1. In a planar coordinate system, the vertical line that runs above and below (north and south of) the origin (0,0). Numbers north of the origin are positive and numbers south of it are negative. 2. In a spherical coordinate system, the y-axis lies in the equatorial plane and passes through 90 degrees east longitude. *See* x-axis, z-axis, Cartesian coordinate system. 3. On a chart, the vertical axis.

y-coordinate *See* x,y and x,y,z coordinates.

Z

z-axis In a spherical coordinate system, the vertical line that runs parallel to the earth's rotation, passing through 90 degrees north latitude, and perpendicular to the equatorial plane, where it crosses the x- and y-axes at the origin (0,0,0). *See* x-axis, y-axis.

zoom To display a larger or smaller region of an on-screen map or image. *See also* pan.

About the Contributors

Andrew A. Beveridge is Professor of Sociology, Queens College and the Graduate School City University of New York. He received his Ph.D. from Yale University. His research includes the impact of location and neighborhood on educational outcome, drug use and drug sales, crime and other factors, as well as the analysis of long-term demographic trends. He has consulted in a number of civil rights cases and consults for *The New York Times,* which has published a series of news reports and maps based upon his analysis of U.S. Census data. Virtually all his work makes extensive use of GIS techniques.

David J. Bodenhamer is Professor of History and Executive Director of The Polis Center at Indiana University-Purdue University, Indianapolis. He earned his Ph.D. degree in history from Indiana University in 1977. A specialist in American constitutional and legal history, he has written and edited six books, including *Fair Trial: Rights of the Accused in American History* and *The Bill of Rights in Modern American: After 200 Years* (with James W. Ely, Jr.), and more than twenty-five articles and book chapters. He also has presented widely on issues related to historical GIS at conferences in the United States, Europe, and Asia.

Peter Collier is Principal Lecturer in Geography at the University of Portsmouth, where he has taught since earning his Ph.D. from the University of Aston in 1980. His current research on the history of cartography focuses on the impact of technological changes in data capture on topographic mapping and the relationship between geography and surveying during the period when geography was becoming established as an academic discipline. Among his many publications are articles on using historical data for environmental studies and military mapping during World War I.

Geoff Cunfer is Assistant Professor of Environmental History and Studies in the Center for Rural and Regional Studies at Southwest State University, Marshall, Minnesota. He explores interactions between people and the natural world, focusing especially on the history of agriculture, agroecology, and land use in the prairies and Great Plains. He received his Ph.D. in history from The University of Texas at Austin in 1999.

Tom Elliott is Director of the Ancient World Mapping Center at the University of North Carolina, Chapel Hill. He holds degrees in computer science, classics, and ancient history, and is completing a dissertation on boundary disputes and Roman imperial administration. He is also Web Information Coordinator for the American Society of Greek and Latin Epigraphy and organizer of the EpiDoc Collaborative, an international effort to develop software- and hardware-independent interchange specifications for scholarly and educational editions of inscribed and incised texts in Greek, Latin, and other ancient languages.

Ian N. Gregory is Research Fellow at the University of Portsmouth. He received an M.Sc. in GIS from Edinburgh University in 1994 and a Ph.D. in geography from London University in 2001. He has published on the use of GIS in historical research in journals such as *Transactions in GIS, Computers Environment and Urban Systems, International Journal of Population Geography,* and *Social Science History,* and he wrote the U.K. Arts and Humanities Data Service's guide to good practice in historical GIS, *A Place in History: A Guide to Using GIS in Historical Research,* forthcoming from Oxbow Books.

Myron P. Gutmann is Director of the Inter-university Consortium for Political and Social Research and Professor of History at the University of Michigan. He earned his Ph.D. in history from Princeton University in 1976. He has published two books and more than fifty articles and chapters, including *War and Rural Life in the Early Modern Low Countries* (1980) and *Towards the Modern Economy: Early Industry in Europe 1500–1800* (1988). His current research is on the relationship between population and environment in the U.S. Great Plains, and on the history of the Hispanic population of the United States in the nineteenth and twentieth centuries.

Trevor M. Harris is Eberly Distinguished Professor of Geography and Chair of the Department of Geology and Geography at West Virginia University. He earned his Ph.D. in geography from the University of Hull, United Kingdom. He has published some forty articles and chapters, is coeditor of a forthcoming book on community participation and GIS, and has received several million dollars in external research funding from major federal and state agencies. Many of his publications focus on the integration of GIS and archaeology. His current research interests are in linking GIS to virtual reality, historical–geographical analysis, societal issues, and environmental impact assessment.

Amy Hillier received her Ph.D. in social welfare from the University of Pennsylvania in 2001. As a Postdoctoral Researcher at the Population Studies Center and Research Associate in the Cartographic Modeling Laboratory, both at the University of Pennsylvania, she is continuing her research on the application of GIS and spatial statistical methods to urban issues, including housing abandonment, housing discrimination, and public health. In addition to expanding her research on historical redlining, she is currently working on a study for the Centers for Disease Control on how neighborhood conditions affect birth outcomes in Philadelphia.

Anne Kelly Knowles is Assistant Professorial Lecturer in Geography at George Washington University. After receiving her Ph.D. in geography from University of Wisconsin–Madison in 1993, she taught geography at the University of Wales, Aberystwyth, and Wellesley College. She has published a dozen essays and one book, *Calvinists Incorporated: Welsh Immigrants on Ohio's Industrial Frontier* (1997), and edited an issue of *Social Science History* on historical GIS (2000). She is now writing a book on labor and technology in the early nineteenth-century iron industry. In fall 2002, she will join the geography department at Middlebury College.

Lewis R. Lancaster is Professor Emeritus of East Asian Languages and Director of the Electronic Cultural Atlas Initiative, based in International and Area Studies, University of California, Berkeley. He received his Ph.D. from the University of Wisconsin in 1968. In recent years he has published widely on Buddhism in China and Korea and has been active in many international efforts to establish standards and guide content development for digital projects in religious history.

David W. Lowe is a historian for Cultural Resources GIS, an office of the National Park Service that assists cultural and historical parks with survey and mapping services. His publications include *Study of Civil War Sites in the Shenandoah Valley* and other resource studies. He served on the staff of the congressionally established Civil War Sites Advisory Commission. He is a founding member of the Civil War Fortifications Study Group and a contributor to the forthcoming park service publication, *Guide to Sustainable Earthworks Management.* He holds an M.A. in American history from George Mason University.

Alastair W. Pearson is Principal Lecturer in Geography at the University of Portsmouth, where he received his Ph.D. in 1996. He has written on the use of geospatial techniques for analyzing historical land use and environmental conditions. Currently, he is preparing an essay on twentieth-century depictions of the third dimension for the Exploratory Essays Initiative, part of the History of Cartography project, and is managing projects on mapping salt marsh change and studying eutrophication in estuaries. He is Assistant Editor and editor of reviews for *The Cartographic Journal.*

Benjamin C. Ray is Professor of Religious Studies at the University of Virginia. He received his Ph.D. in history of religions from the Divinity School, University of Chicago, in 1971. His work on turning to technology to enhance classroom teaching led to a collaborative research project on the Salem witch trials, supported by Virginia's Institute for Advanced Technology in the Humanities and the National Endowment for the Humanities. Important lessons learned involve interdisciplinary collaboration, grasping the potentialities of computer technology, and widening the educational purposes of academic research.

David Rumsey is President of Cartography Associates in San Francisco, a company that specializes in digital publishing. Rumsey received B.A., B.F.A., and M.F.A. degrees from Yale University, where he was Lecturer in Art from 1969–1973. While working in real estate development and finance he formed the David Rumsey Historical Map Collection, which numbers over 150,000 historical maps and is one of the largest private map collections in the United States. His online map library, www.davidrumsey.com, features more than sixty-five hundred high-resolution images from the collection, mainly maps from North and South America, and has recently added an online GIS browser.

Aaron C. Sheehan-Dean is a doctoral candidate in history at the University of Virginia, where he is also a GIS specialist with the Virginia Center for Digital History. He is currently writing his dissertation on the social and political experience of Virginia's Confederate soldiers during the U.S. Civil War.

Humphrey R. Southall received his Ph.D. in geography from Cambridge University in 1984. He is Reader in Geography at the University of Portsmouth and Director of the Great Britain Historical GIS Project. That project grew out of his research into the history of Britain's north–south divide and spatial divisions of labor. He has also published extensively on the historical geography of trade unionism and strikes and on the contribution of trade unions to the development of Britain's welfare state.

Richard Talbert is William Rand Kenan Jr. Professor of History and Classics at the University of North Carolina, Chapel Hill, and current President of the Association of Ancient Historians. He earned a Ph.D. in Classics from Cambridge University in 1972. He wrote *The Senate of Imperial Rome* (1984), which won the American Philological Association's Goodwin Award of Merit, and edited the *Barrington Atlas of the Greek and Roman World* (2000). His current research concentrates on Roman roads, maps, and worldview.

Meredith Williams is the GIS Manager at Stanford University. She assists faculty and students from all departments with using GIS as a research tool. She studied geography and environmental studies at UCLA, where she conducted research in forest ecology. Currently, Meredith is involved in a project using remote sensing and GIS to study land-use change in the Red River Delta of Vietnam. In another project, she is using GIS to model earthquake scenarios and emergency response vehicle routing in the San Francisco Bay area. She is interested in the application of GIS and remote sensing for natural hazards management.

Book design, production, and copyediting by Michael Hyatt

Cover design by Amaree Israngkura

Image editing by Jennifer Johnston

Printing coordination by Cliff Crabbe